Optics in Photography

Cover: photograph of a cutaway section of a Vivitar Series 1 90mm *f*/2.5 lens

Optics in Photography

Rudolf Kingslake

SPIE OPTICAL ENGINEERING PRESS

A publication of SPIE—The International Society for Optical Engineering
Bellingham, Washington USA

Library of Congress Cataloging-in-Publication Data

Kingslake, Rudolf.
Optics in photography / Rudolf Kingslake.
p. cm.
"A Publication of SPIE—the International Society for Optical Engineering."
Includes bibliographical references and index.
ISBN 0-8194-0763-1
1. Photographic optics. I. Title.
TR220.K56 1992 92-11861
771.3'5--dc20 CIP

Published by SPIE—The International Society for Optical Engineering
P.O. Box 10
Bellingham, Washington 98227-0010

Design: Matt Treat
Composition: Carrie Binschus

10 9 8 7 6 5 4

Printed in the United States of America

Table of Contents

Preface

I am greatly indebted to the Society of Photo-Optical Instrumentation Engineers (SPIE) for undertaking the publication of this book. It is really an updated third edition of my previous book entitled *Lenses in Photography*, two editions of which were published, in 1951 and 1963, and which have long been out of print. The new title is more truly representative of the contents, as such subjects as scene brightness, emulsion speed, exposure determination, and flash equipment do not involve lenses at all. However, the rest of the book is concerned with lenses, including perspective considerations, depth of field, resolving power, viewfinders, image brightness, enlargers, projectors, and stereoscopy.

As was stated in the two previous editions, this book is addressed mainly to the advanced amateur photographer who wishes to know more about the equipment and how it should be used. No consideration has been given to the purely professional aspects of the subject, such as 35mm motion pictures and sound recording, medical and x-ray photography, aerial photographs, photogrammetry, astronomical and space photography, high-speed cameras, photocopying machines, and halftone printing. Nor has any attention been paid to the design and construction of modern cameras and their electronic components, or to the chemistry of the photographic process.

The principal changes in the present edition are the result of the rapid advance in the complexity of the modern amateur camera, particularly the SLR and rangefinder models using 35mm film. Interest in 8mm and 16mm motion pictures has waned considerably, and those now interested in making movies are more likely to be using a camcorder than a film camera. Other major changes are the shift from flashbulbs to electronic flash, the general use of multicontrast printing paper, a strong interest in underwater photography, and the introduction of cameras equipped with automatic focusing. The modern camera, full of mechanical and electronic gadgetry, is a remarkable piece of equipment, and the user is tempted to rely entirely on the automatic features, which it must be admitted generally work perfectly.

Chapter 7 on lens types has been entirely rewritten, as the whole field has changed. Very few of the old classical lenses are still in production, while zoom lenses and reversed telephotos are becoming almost universal. Large cameras such as the Speed Graphic and other 4 × 5 view cameras have virtually disappeared, except for those being used by the few individuals who still cling to them. The possession of a named lens, formerly the mark of a serious photographer, is no longer applicable, as lenses today do not have names, and indeed it is hard, if not impossible, to ascertain the structure of any modern lens.

Aside from changes in cameras, the external form of slide projectors has changed from the older tall models to those of a flatter shape, thanks to the introduction of projection lamps that can be operated in a horizontal position. The familiar overhead projector, or Vuegraph, has largely replaced slides as a lecture medium. Color photography has almost entirely replaced the older black-and-white pictures among amateurs, and in magazines and even in some newspapers. Stereoscopy has, however, remained unchanged except for a brief introduction of multilens cameras that yield a ribbed lenticular print.

I am indebted to various sources, acknowledged individually in the first edition of *Lenses in Photography*, for many of the illustrations used in this book.

My sincere thanks are due to Eric Pepper of the SPIE staff, who supervised the production of this book with great care and patience.

Rudolf Kingslake *March 1992*
Rochester, New York

Optics in Photography

Chapter 1

Perspective

The Center of Perspective

When we use photography to make a permanent record of something we have seen, we are endeavoring to represent a three-dimensional aggregate of objects upon a plane surface. Thus, a photograph is a point projection of a three-dimensional scene, commonly referred to as a "perspective" view, with the "center of perspective" at the camera lens itself.

Optically, our eye is nothing but a camera, and when we look at a scene with one eye open we actually see the same kind of perspective as is recorded by a camera. We could, if we wished, plot out a picture of an assemblage of objects upon a sheet of glass erected between our eye and the objects, as indicated in Fig. 1.1. In this diagram, the eye is supposed to be at *P*, and the appearance of the house as seen by this observer is shown traced on the sheet of glass. The only difference between this glass-plate picture and the picture formed by a pinhole camera is that in the camera

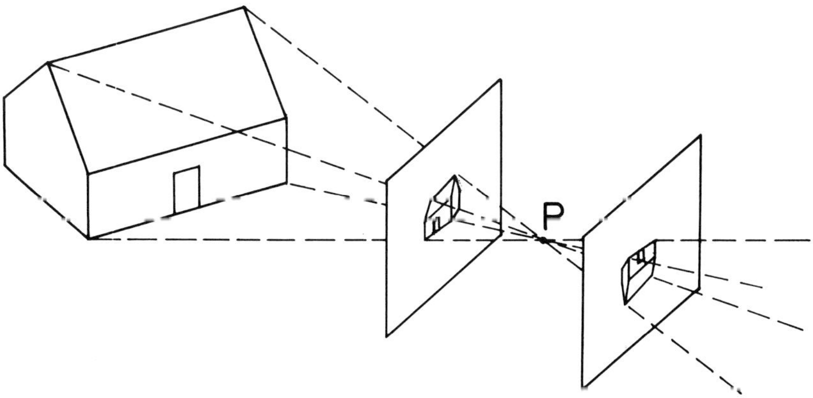

Figure 1.1. The meaning of "center of perspective."

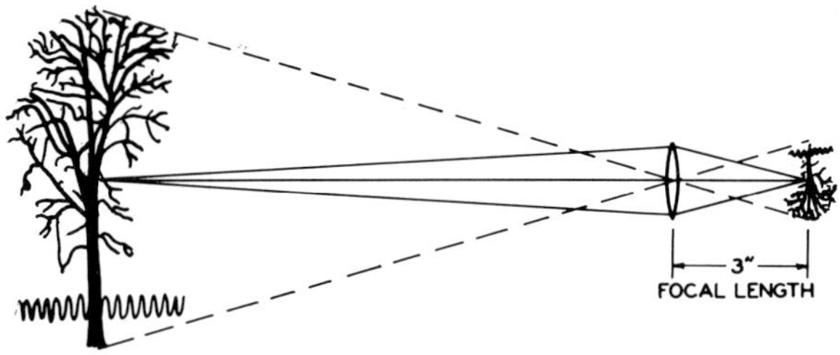

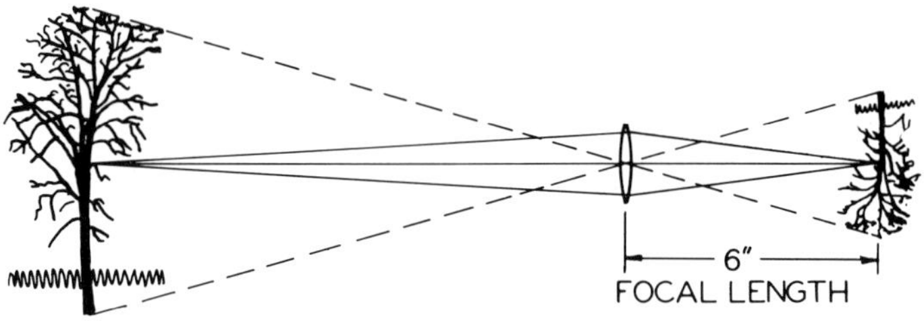

Figure 1.2. A lens of long focus produces a larger image than one of short focus.

the rays cross each other at the pinhole (at *P* in the diagram) and are intercepted by a sensitive film placed behind it. The inverted picture projected on the film will be identical with that traced on the glass plate, provided the film and the glass plate are parallel to each other and are both the same distance from *P*. The point *P* is then the "center of perspective" of the photographed picture. If a lens is used instead of a pinhole, the entrance pupil* of the lens is the center of perspective, and everything that is said here on the subject of perspective applies equally well to the pinhole or to the lens camera. A photograph is thus a two-dimensional rendering of a three-dimensional object, the projection lines all passing through a common center. The size of the projected image will depend on the distance from the center of projection to the film plane, but all images on parallel film planes projected through the same perspective center will be geometrically similar.

We thus reach the important conclusion that we shall obtain the same view of a three-dimensional object or group of objects with any camera whatsoever, provided that the position of the camera lens is fixed and the film plane is in a fixed direction. A lens of long focus will merely produce a larger picture than a lens of shorter focus (Fig. 1.2).

*The entrance pupil is the image of the diaphragm as seen from the front of the lens.

Conventions in Perspective

For many centuries, it has been the established custom to draw or paint a picture as if it were projected upon a vertical plane, and in photography we ordinarily try to follow this rule. If by accident or design we hold the camera during exposure in such a way that the film plane is not vertical, then vertical lines in the object will appear to lean toward each other in the photograph (keystone distortion). The effects of tilting the camera upward or downward are shown in Fig. 1.3 and Fig. 1.4, respectively. This is not a defect of the photographic process, for the camera is indeed recording faithfully what we would see if we looked with one eye in the same direction as the camera was pointing. The difference is that when we see lines apparently converging but known to be parallel, we unconsciously interpret them as being parallel; indeed, we can actually detect the presence of

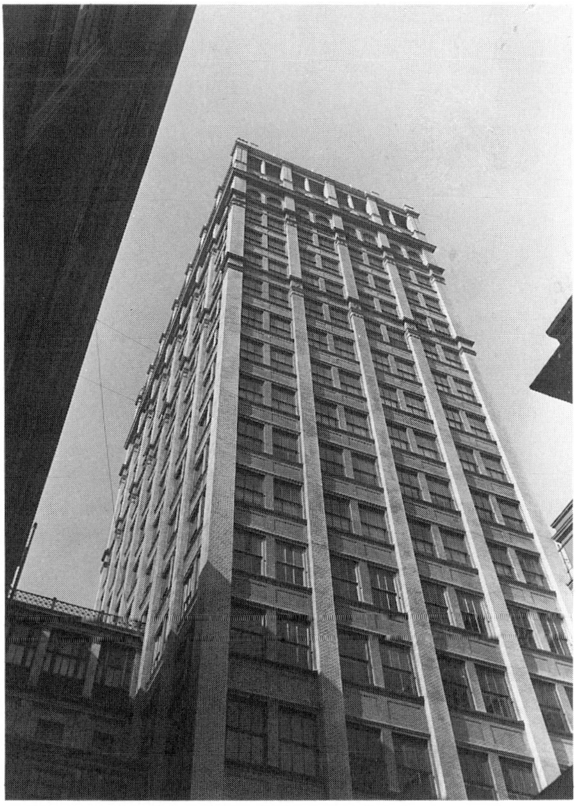

Figure 1.3. Keystone distortion resulting from an upward tilt of the camera.

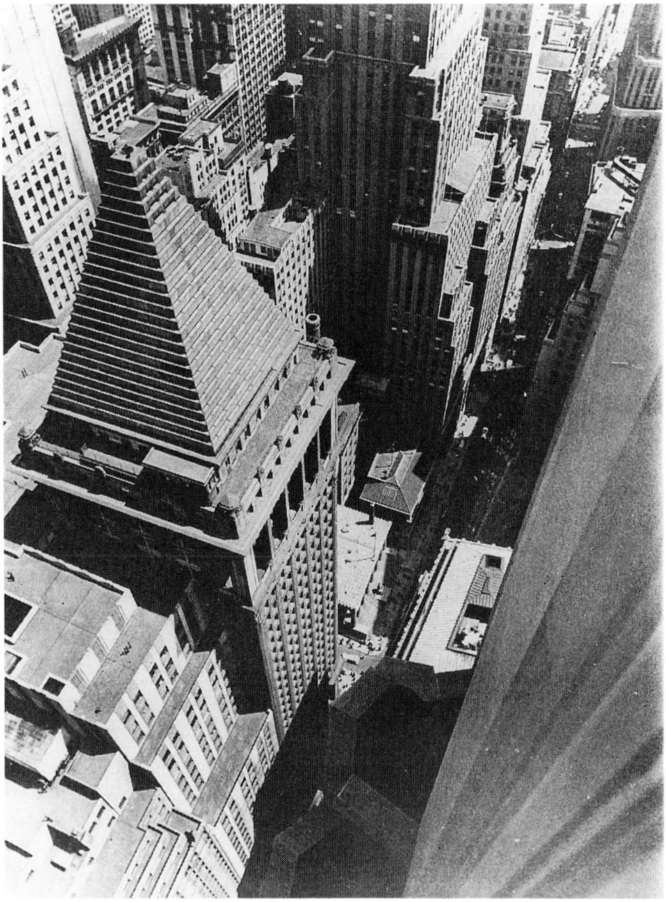

Figure 1.4. Keystone distortion resulting from a downward tilt of the camera.

a small amount of genuine convergence if it exists. However, we do not object to a *sideways* tilt of a camera, causing horizontal parallel lines to converge (Fig. 1.5), probably because this corresponds simply to a change in the horizontal direction of view, our most common visual experience. In printing such a photograph, great care must be taken to keep vertical lines vertical on the print.

It appears that an upward or downward tilt to a camera is undesirable only if it is present to a moderate extent, but very little objection is raised to converging vertical lines if the upward or downward tilt is greater than about 30°. In photographing tall buildings, a strong convergence is a commonly used device to suggest height.

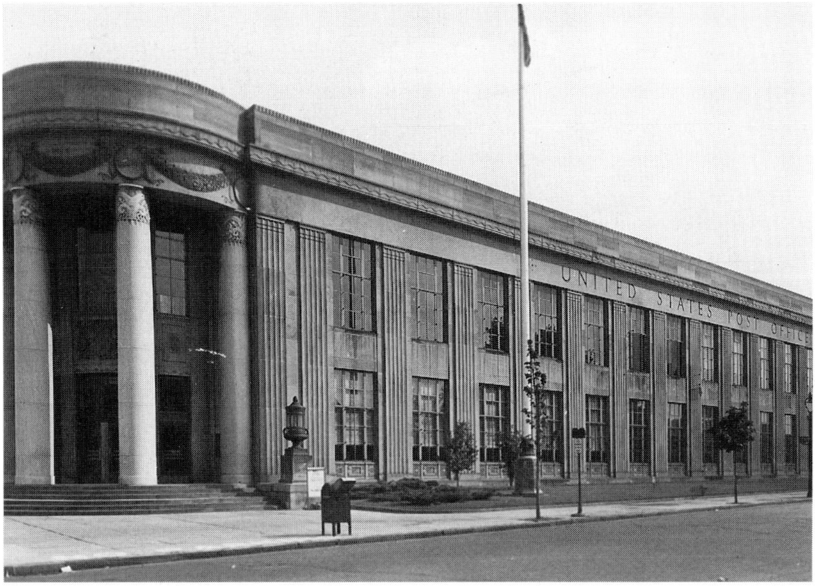

Figure 1.5. The effect of tilting a camera sideways.

If it is necessary to photograph a tall building from the ground without converging verticals, we can either rectify the convergence by use of a tilted enlarger (see page 203) or we can use a camera equipped with a "rising front." This is a device for elevating the lens, and since the image moves with the lens, the top of the tall building will be brought onto the film without the necessity of tilting the camera. In this case we must use a lens that will give good definition over the greatest angle required, namely, θ in Fig. 1.6. Many lenses will not do this unless stopped down very

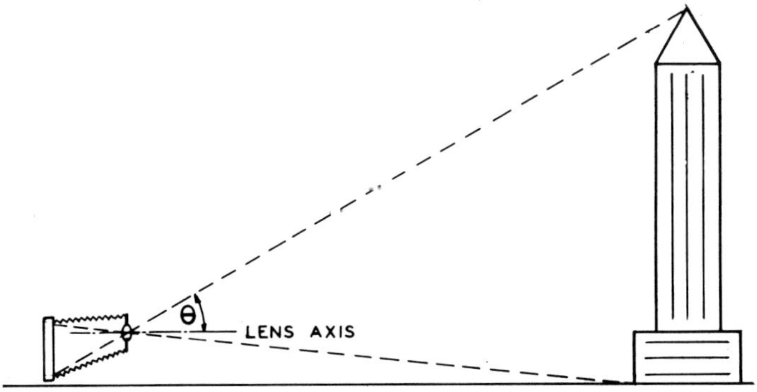

Figure 1.6. Illustrating the use of a rising front in photographing tall buildings.

considerably. To carry the procedure still further, a "swing back" is sometimes used; the essential requirement is, however, to keep the planes of object and film parallel to each other. A typical camera for this purpose is shown in Fig. 1.7.

Another useful rule for satisfactory perspective is that we should not include too wide an angular field of view in one picture. If we stand in a room and let our eyes roam over the walls and ceiling, we realize that we are looking critically at only a small part of the room at any one time. Indeed, the eye can see clearly over a half-angle of perhaps 10° from the center of the field. We can see large objects, and movements of any kind, over an angle of perhaps 25° from the center of the field. To see a wider angle than 25° requires that we roll our eyes consciously and scan the objects one at a time.

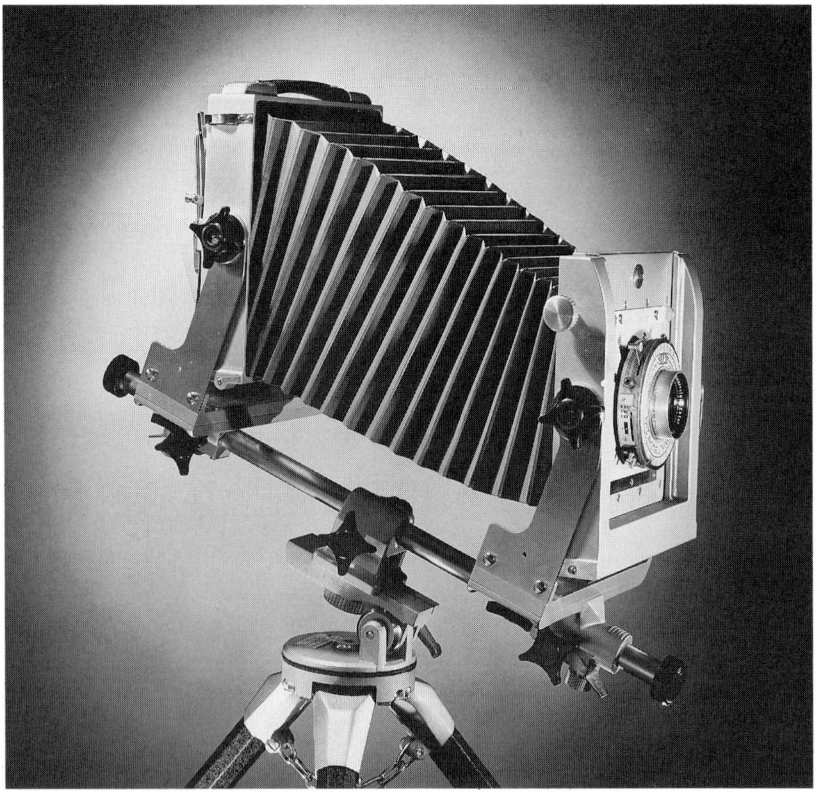

Figure 1.7. The Calumet view camera.

By far the most important rule for correct perspective in photography is that the final print must be viewed from approximately its correct center of perspective, so that the angles subtended at the eye by the various images in the picture will be the same as the subtense angles of the original objects at the camera lens. For contact prints, the center of perspective corresponds to the actual location of the lens in the camera, opposite the middle of the picture and distant from it by the focal length of the camera lens. For enlargements, the distance of the center of perspective from the print is found by multiplying the focal length of the camera lens by the enlargement ratio. Thus, for a negative made in a 35mm camera with a 50 mm (2-inch) lens, and enlarged 10 times in printing, the center of perspective is at 20 inches from the print, and the picture should be viewed from this point. The gain in realism obtained by enlarging small negatives in this way is quite marked and often astonishing.

The lateral position of the eye in relation to the center of perspective is also important. This fact explains the serious distortion that results when we look at a motion-picture screen from the end of the front row of seats, the center of perspective being actually located on a line joining the projector to the screen. In planning a large mural, which is to be viewed from the floor of a room, it is advisable to have the camera low and use the rising front. The opposite effect, with the camera looking down on the subject, would be very unpleasant in such a case.

The Field Covered by a Lens

Every lens projects light onto a circular field that is limited in size by the vignetting or cutting of oblique light by the lens barrel. However, in very few lenses is the definition sharp to the extreme limit of this circle of illumination. Since good definition is required in any practical application of the lens, it is customary to state the field of a lens in terms of the angle over which good definition is obtainable (Fig. 1.8). This angle generally increases somewhat as the lens is stopped down to a smaller aperture.

Since most photographs are taken on a square or rectangular film area, it is necessary that the film format should fit into the circle of good definition of the lens. Thus, the diameter of this circle must be equal to, or greater than, the diagonal of the film.

The "Normal" Focal Length for a Camera Lens

For ordinary photography, the "normal" field is usually such that the diagonal of the negative is equal to the focal length of the taking lens. This

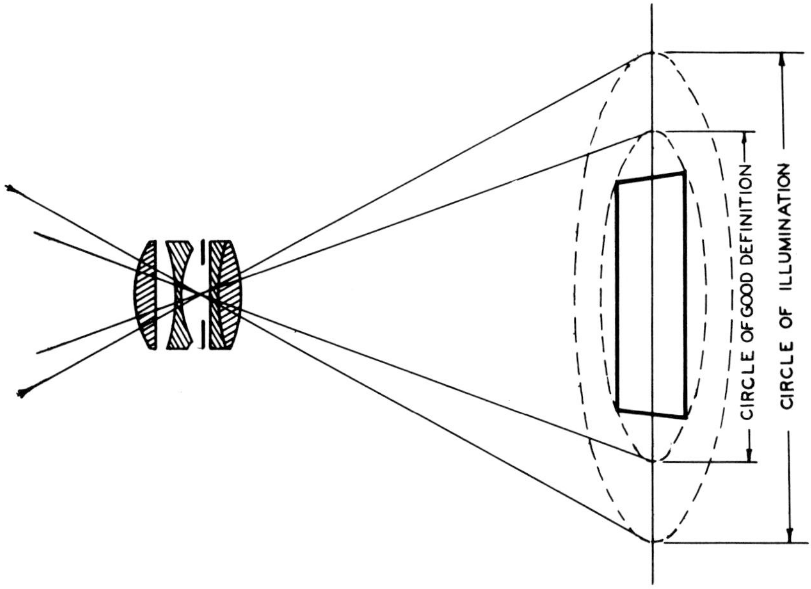

Figure 1.8. The circle of illumination and the circle of definition of a lens.

computes to be a total field of 53°, or a half-field of 26.5°. As has been mentioned, this angle is rather wider than the eye can cover at a glance, but in practice we tend to view most photographs from a point well beyond the center of perspective, and we unconsciously scan a print with our eyes. Such an angle of view is therefore not objectionable. Many photographic prints, too, are cropped in printing so that the whole of the lens field is not recorded.

A *wide-field* lens will cover an angular semifield of about 30° to 35°, and a true *wide-angle* lens will cover a semifield of 45° to 50° (see Fig. 1.9). Hence, a given film format will be covered adequately by a wide-angle lens having a focal length equal to about half the picture diagonal. Naturally, the field covered by the camera will not be increased by using a wide-angle lens of the same focal length as the normal lens; we can gain field only by the use of a shorter-than-normal focal length.

Some *narrow-angle* lenses are loosely called "telephoto" lenses because they have a longer focal length than the normal lens and thus give a picture to a larger scale. However, the name "telephoto" should be restricted to a lens of a particularly compact type of construction (see page 148), in which the distance from the front of the lens to the film plane is

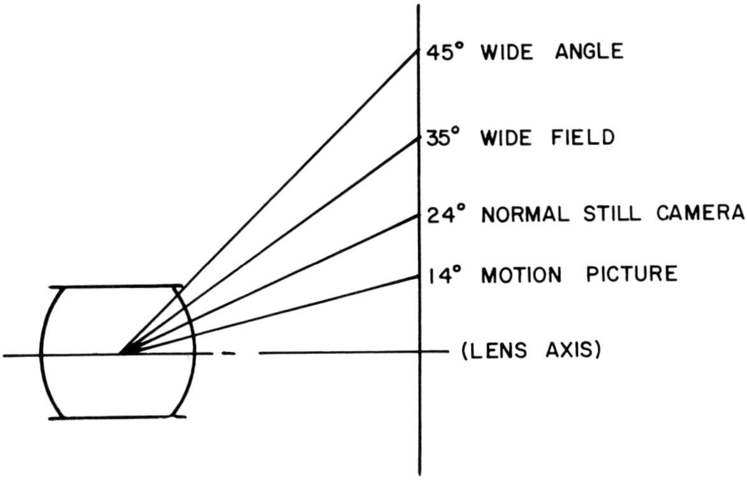

Figure 1.9. Some typical angular fields of lenses.

less than the focal length of the lens.

In motion-picture photography it has always been customary to use relatively long-focus lenses, the "normal" angular semifield being only about 14°. This was probably done originally to keep the camera well away from the actors in order to give them greater freedom of movement along the line of sight. If the camera were close to the subject, with a short-focus lens, the apparent size of the actors would appear to grow or shrink rapidly as they moved, and they might even walk completely out of focus in a couple of steps! Moreover, the "panning distortion" mentioned on page 20 is less noticeable if a long-focus lens is used. The use of a narrow angular field greatly helps the lens designer to make wide-aperture lenses of good quality, which are very necessary with motion-picture cameras having a fixed exposure time.

The matter of perspective must not be overlooked in this connection, as it is generally desirable that the center of perspective should fall at about the middle of the audience in a theatre. Hence the angular field of the camera should be about twice that of the projector, which again brings the camera semifield to about 14°. A wide-angle movie lens then covers a half-angle of about 20°, which is actually less than the field of a normal lens in still photography.

The following table of picture diagonals for some standard film sizes may be of interest. They are taken from ANSI Standard PH3.501-1987.

(a) Still cameras

Film name	Negative area	Diagonal (mm)	Diagonal (inch)
Disc and Minox	8 × 11 mm	13.6	0.53
110	13 × 17 mm	21.4	0.84
half 135	17½ × 24 mm	30.1	1.19
126	28 × 28½ mm	40.0	1.57
135	24 × 36 mm	44.0	1.73
828	28 × 40 mm	48.8	1.92
Sq. 127 (1⅝ × 1⅝ in.)	41 × 41 mm	58.0	2.28
127 (1⅝ × 2¼ in.)	41 × 56 mm	69.4	2.73
Sq. 120 (2¼ × 2¼ in.)	56 × 56 mm	79.2	3.12
120 (2¼ × 3¼ in.)	56 × 82½ mm	99.7	3.93
116	2½ × 4¼ in.	125.2	4.93
¼-plate	3¼ × 4¼ in.	135.9	5.35
	4 × 5 in.	162.6	6.40
	5 × 7 in.	218.5	8.60

(b) Motion-picture cameras

Size	Camera Frame (mm)	Camera Diagonal (mm)	Projector Gate (mm)	Projector Diagonal (mm)	Normal focal length of camera lens
35mm silent	19.05 × 25.37	31.75	17.26 × 23.01	28.76	2 inch (50 mm)
35mm sound	16.03 × 22.05	27.26	15.24 × 20.96	25.91	2 inch (50 mm)
16mm	7.42 × 10.22	12.63	7.21 × 9.65	12.05	1 inch (25 mm)
Super-8	4.22 × 5.77	7.15	4.01 × 5.36	6.69	½ inch (13 mm)
8mm	3.51 × 4.80	5.95	3.28 × 4.37	5.46	½ inch (13 mm)

"True" and "Apparent" Perspective

From the preceding discussion it should be clear that if we look at a photograph from some point other than its true center of perspective, we must expect to see a distorted representation of the original scene. For instance, if our eyes are considerably too far away from the picture, foreground objects will appear too large, and background objects relatively too small. This effect is particularly noticeable in photographs taken with a wide-angle lens, such as that in Fig. 1.10. The center of perspective of the lower photograph is at about 6 inches from the print, and if our eye is placed there, we have the impression that we are looking at a car from a reasonable distance away. For the upper picture, a wide-angle lens was used and the camera was moved very close to the car. The center of perspective of the

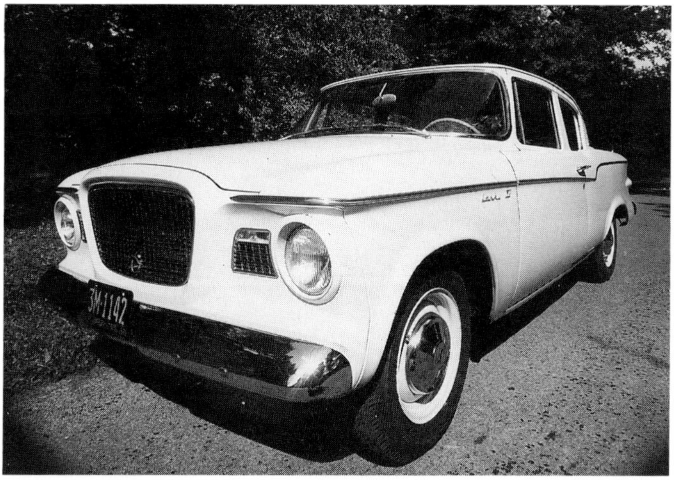

(a)

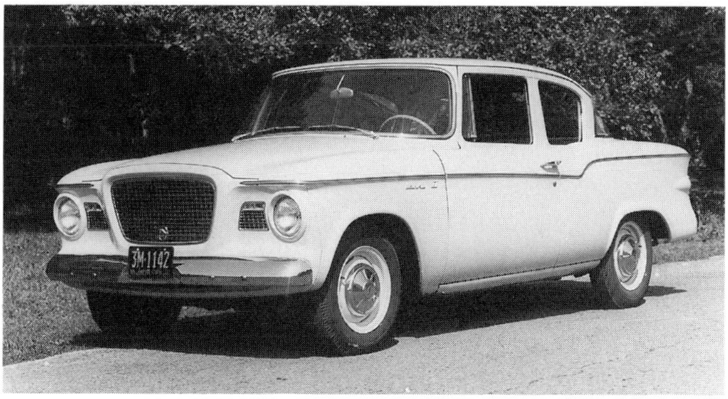

(b)

Figure 1.10. Perspective distortion. The distance of the camera from the front wheel was five feet in (a) and 25 feet in (b).

print is now only some 2 inches away, and if we place our eye there, we again see a reasonable picture of a car standing quite close to us. However, if we view the upper print from 12 to 15 inches away, the foreground becomes apparently magnified, and the near front wheel appears absurdly large.

The absurdity in this case is made up of two contributory factors: First, the *true perspective* of the picture is unusual in that the camera was placed decidedly close to one wheel of the car, whereas a better proportioned view would have been obtained by moving the camera much further away. Second, the *apparent perspective* of the picture is distorted by the observer since he or she is looking at it from a point far beyond its center of perspective. Unfortunately, both of these factors act in the same direction, with the result that few wide-angle photographs are really fair reproductions of the original scene.

We thus find that there are two independent ways of varying the perspective in a photograph: (1) by moving the camera lens relative to the original object, and (2) by moving the eye relative to the finished print. The first movement controls the true perspective in the photograph, which is fixed once the picture has been taken. Changing the focal length of the camera lens or enlarging the negative during printing cannot affect this perspective in any way. The second movement changes the apparent perspective of the photograph and thus indirectly affects the ultimate appearance of the scene. For the perspective to be correct, the photographic print must appear to the observer just as the original scene appeared to the photographer. This will occur only if the print is viewed from its correct center of perspective. An extreme example of the effect of camera position on the true perspective of a scene is shown in Fig. 1.11.

It should be added that if we assume that all prints are to be viewed at a fixed distance, say 15 inches from the observer's eye, then a change in either the focal length of the camera lens or in the enlarger magnification will affect the apparent perspective, because either of these changes will move the center of perspective of the print. Neither change, of course, will affect the true perspective in the picture itself since this depends only on the original position of the camera.

A minor effect of viewing a picture from too great a distance is that the depth of the scene, from the furthest to the nearest objects in it, appears to be increased, while if we look at the picture from too close a distance the scene appears collapsed and too shallow. This remark applies particularly to a large mural, or even to an oil painting, where the center of perspective may be many feet away from the picture, and the viewer can easily be too close to it. Indeed, if the average viewing distance is known in advance, the perspective conditions can be arranged so that the picture will enjoy the maximum degree of realism.

Disproportionate Magnification of the Foreground

When we look at a picture, we instinctively assume that the camera was located where our eye happens to be. Consequently, if we look at a close-up picture such as the upper view in Fig. 1.10 from too far away, we instinctively assume that the front wheel of the car was really too large, whereas if we look at it from the center of perspective, we get the correct

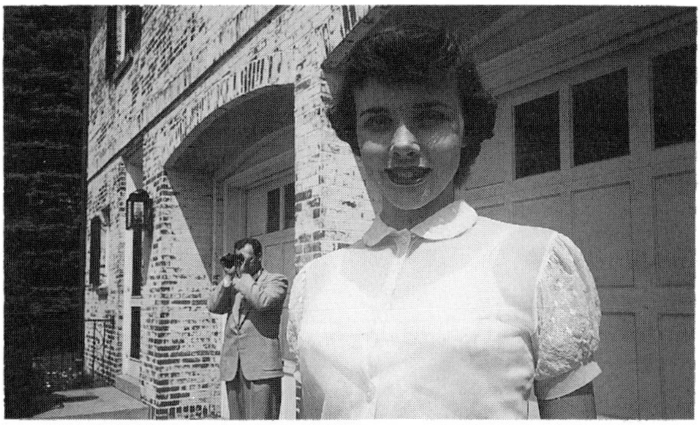

(a)

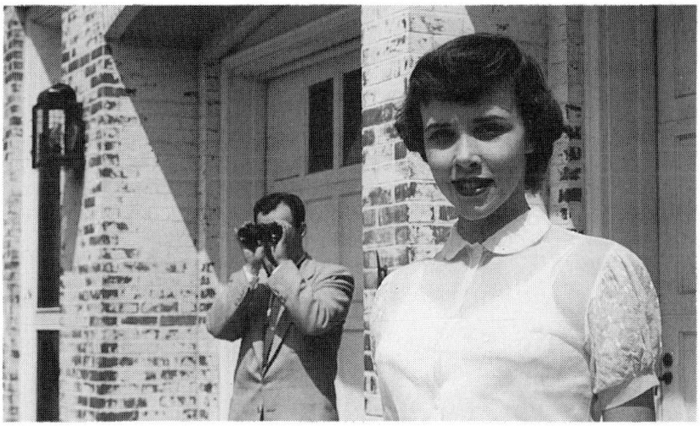

(b)

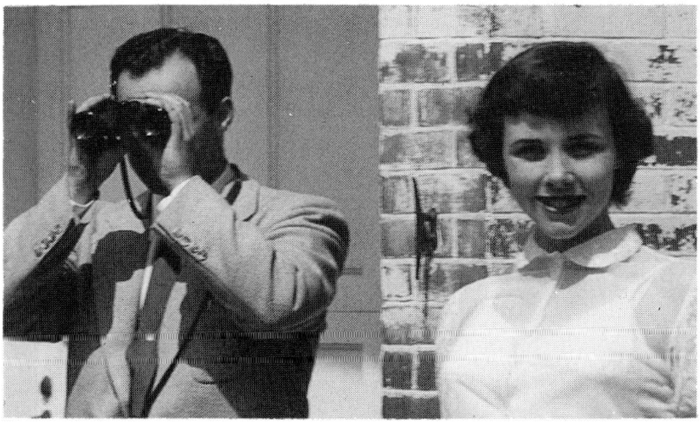

(c)

Figure 1.11. The effect of camera position in relation to a scene. For (a) the camera was placed at 2 feet from the girl; for (b) it was at 6 feet; and for (c) it was moved back to a point 38 feet from the girl. The negatives were enlarged in printing so as to make the girl appear at the same size in each print.

impression that the camera was too near the car when the photograph was made. This effect may be better understood by a consideration of Fig. 1.12.

Suppose *A* and *B* are two objects being photographed. If the camera lens is at M, the photograph will have the proportions shown at X, the height *aX* being 1.46 times the height *bX*. This represents a wide-angle view. On the other hand, if a normal lens is used at *N*, the photograph will look like Y, and the height *a'*Y is now 2.0 times the height *b'*Y.

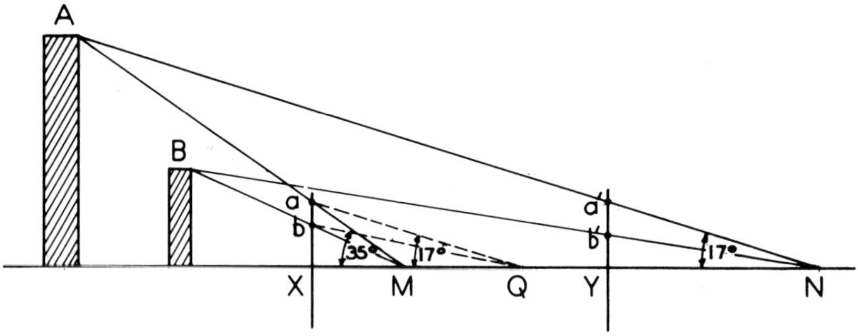

Figure 1.12. Perspective distortion when a picture is viewed from a point beyond its center of perspective.

Obviously, the correct viewpoint for the picture X is M, and if we view X from a point considerably too far away (say Q) so that the height *aX* subtends only 17° instead of the correct angle of 35°, the object *B* will appear too high, since for a 17° angle of view, the correct relative height of *B* compared with A would be 0.5 (as in picture *Y*). The reverse effect is also seen; namely, if we take a picture with a long-focus lens and look at it from a point too close to the print, we shall see background objects relatively too large and foreground objects too small. However, this defect is easily remedied by moving well back from the print; it is much harder to improve a wide-angle view by going closer to it because this involves a strain of the accommodation, and it requires the eye to roll around to scan the picture instead of taking in the whole scene at a single glance. Consequently, very few people are conscious of the "bad" perspective produced by a long-focus lens, but it is common to hear complaints of the "distortion" resulting from the use of a wide-angle lens.

The Use of a Magnifier to View a Print

If we desire to view a small print in which the center of perspective is but a few inches away, it is a great help to use a magnifying lens having a focal length approximately equal to the distance of the perspective center from the print. This enables the eye to relax its accommodation and still see the view clearly from its correct center of perspective. Hand viewers for color slides are constructed on this principle, the focal length of the viewing lens being generally 2½ to 3 inches, which is entirely satisfactory for slides made with the normal 2-inch lens, and it is moderately satisfactory for slides taken with the other interchangeable lenses.

Viewing a Print with Both Eyes

In all the foregoing remarks, it is assumed that a single eye is used to view the print. This is generally not the case, and we must now consider briefly the perspective effects due to the use of binocular vision.

Our photograph is a plane representation of a three-dimensional aggregate of objects, which in most cases were at a much greater distance from the camera than the final print is from the observer. Thus, when looking at a print, we shall need to exert our accommodation and converge the axes of our eyes far more than would be necessary were we looking at the original object from the camera position. This tends to give an unnatural appearance, and we feel that we are looking at a model or scaled down reproduction of a scene and not the actual scene itself. However, if the scene is projected optically upon a fairly distant screen, or if we view the print through a large lens having the print at its focal plane, then we relax our accommodation and convergence, and the scene instantly becomes "real" and much more satisfying.

Other Perspective Distortions Caused by a Wide-Angle Lens

In addition to the apparent enlargement of the foreground that has already been discussed, the use of a wide-angle lens may lead to some other types of perspective distortion. For example, if we photograph a row of equally broad *flat* objects on a flat film, the images will also be equally broad provided the lens is free from aberrations. However, if a row of equal *spheres* is photographed, the images will be broader at the ends of the row than in the middle of it, and will appear as horizontal ellipses. In Fig. 1.13 the lines *AA* represent rays from the outer sides of the spheres, whereas *BB* represent the rays from the sides of the flat objects. As all the *BB*'s are projected to the same width on the film, the images of *AA* will be wider at the ends of the picture than in the middle. Viewing such a wide-angle photograph from its center of perspective helps to remove this distortion effect.

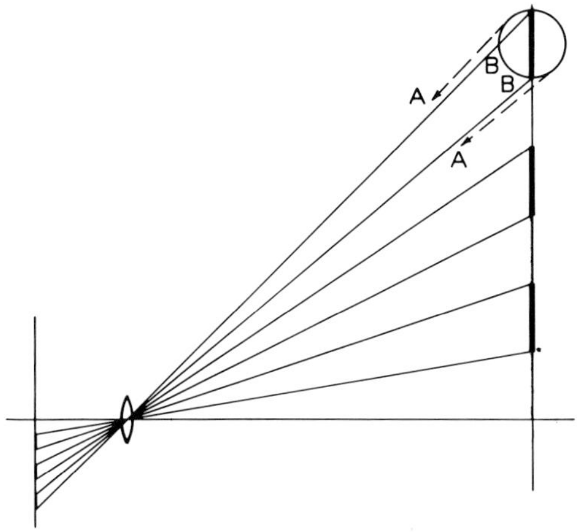

Figure 1.13. Illustrates the apparent broadening of spherical objects at the ends of a wide-angle photograph.

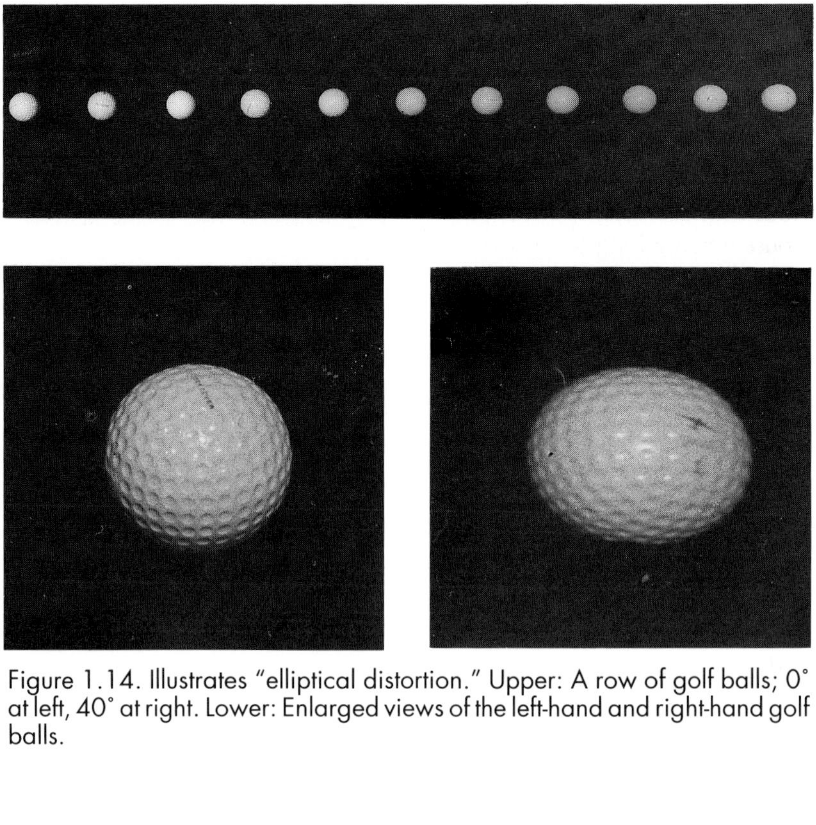

Figure 1.14. Illustrates "elliptical distortion." Upper: A row of golf balls; 0° at left, 40° at right. Lower: Enlarged views of the left-hand and right-hand golf balls.

To illustrate this form of wide-angle distortion, a row of golf balls was photographed with a wide-angle lens (Fig. 1.14). It is seen at once that the balls at the extreme field angles appear elliptical, their radial diameter being equal to their true diameter divided by the cosine of the field angle, in this case 40° from the axis. The cosine of 40° is 0.766, its reciprocal being 1.30; hence the golf ball at the most extreme field angle appears 1.30 times as wide as it is high, as shown in this picture.

This phenomenon is noticeable in long group photographs taken with a wide-angle lens on a flat film, in which the persons standing at the extreme ends of the group appear to be broadened considerably in a lateral direction. The heights of the persons are, of course, not affected by this type of distortion. In banquet photographs taken with a camera high up in one corner of the room, this radial broadening effect leads to some curious elliptic distortions in photographs of persons sitting close to the camera.

Perspective Distortions Caused by Long-Focus Lenses

As has been remarked, these do not usually obtrude sufficiently to cause comment, but in some cases, such as when using a long-focus lens to take movies of a distant object travelling directly toward or away from the camera, a peculiar collapse of space occurs and a horse or a runner, for example, will seem to be making a tremendous effort with no apparent change in position. A train moving toward the camera will appear almost stationary in such a case, because the distance covered by the train is small compared to its distance from the camera. A person standing in front of a house but in reality quite distant from it, will appear to be close up to the house when a long-focus lens is used. Similar effects are frequently seen when looking through a pair of binoculars.

Perspective in Zoom and Dolly Shots

In motion picture photography, it is a common practice to move the camera on a "dolly" toward or away from the subject during the taking of a scene to vary the image size in a continuous manner. An alternative procedure is to use a zoom lens in which the focal length can be continuously varied by turning a lever or crank.

If all the objects in the scene lie in one plane, then a dolly shot and a zoom shot will appear indistinguishable to the audience, but if objects at various distances are involved, these two types of shot will appear quite different. Suppose an actor is standing in front of a distant mountain. Then if the camera is slowly moved toward the actor, his size on the film will increase, but the size of the distant mountain will remain unchanged.

However, if a zoom lens is used, the person and the mountain will both increase in size at the same rate, which is a decidedly unnatural effect.

Perspective in Projected Images

Because of the enormous popularity of optical projection, both for motion and still pictures, it is necessary to consider in some detail the perspective relationships involved.

Suppose the original view subtended an angle θ at the camera lens (Fig. 1.15); then the correct viewpoint for the projected picture must be such as to maintain this angle θ at the observer's eye. The center of perspective B of the projected picture will therefore divide the distance from the projection lens to the screen in the same proportion as the original camera point A divides the distance from the film to the projection lens. For ordinary conditions of projection, this latter proportion is practically equal to the ratio of the focal length of the camera lens to that of the projection lens.

As most projectors use a lens having a focal length about 2 to 2½ times the focal length of the lens normally used on the camera, for most projected pictures (contact prints or reversal transparencies) the center of perspective will be about midway between the projector and the screen. It is desirable, therefore, that the audience be much closer to the screen than the projector if true perspective reproduction is to be obtained.

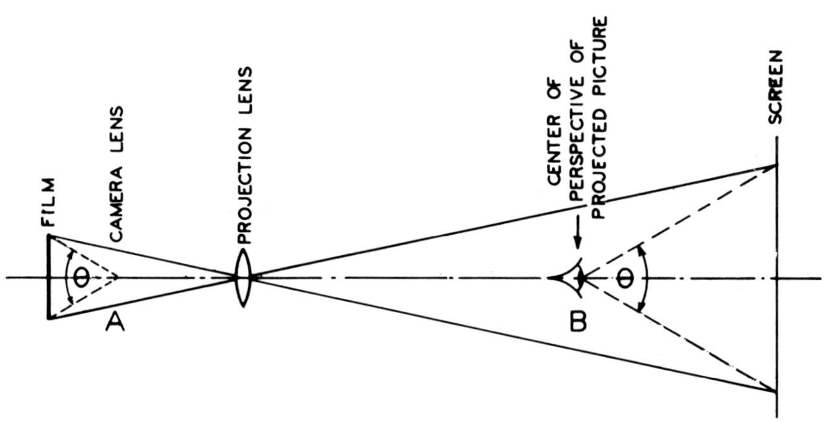

Figure 1.15. Perspective in projected images.

The Use of Interchangeable Camera Lenses

The apparent perspective of projected pictures (motion or still) is thus affected very much by the focal length of the camera lens, even though the true perspective depends only on the position of the camera. If long-focus lenses are used, the center of perspective moves back farther from the screen and it may fall behind the projector entirely, e.g., in 16 mm motion pictures taken with a 3-inch lens and projected with a 2-inch lens. If the observer remains at his normal position midway between projector and screen, he will be situated much closer to the picture than the center of perspective, with the result that foreground objects will appear relatively too small and distant objects too large. The reverse effect is observed in pictures taken with a wide-angle lens.

In spite of this, interchangeable lenses are often useful. It is never desirable to enlarge a small portion of a negative excessively, because of loss of definition, graininess, and the presence of unavoidable tiny marks or scratches on the negative that become conspicuous when greatly enlarged. Moreover, motion pictures and color slides are invariably projected from the original picture size with no opportunity to enlarge or reframe the picture to improve the composition. Having decided on the most desirable viewpoint, therefore, the focal length of the lens should be chosen so as to fill the available film area as fully as possible with the desired subject matter.

A little consideration is often necessary before selecting the best position for the camera. For example, in taking movies of a football game, if a short-focus lens is used and the camera is situated close to the edge of the field, the near players will more than fill the picture whereas distant players will be much too small. It is far better to use a long-focus lens from a point high up on the grandstand so that the relative changes in distance of the players will be less.

Change of Image Size When a Camera is Rotated about a Vertical Axis

There is a plane in the object space, perpendicular to the lens axis, that is an image of the film surface, and all objects lying in this plane will be reproduced at the same magnification. An object lying closer than this focused plane will appear too large, while an object located beyond that plane will appear too small. This argument can be used to explain keystone distortion.

Thus, when a camera is "panned" around a vertical axis, any object that is on the lens axis at one camera position will be too close to the camera after panning, so the object will appear to grow in size while the camera is panned (Fig. 1.16). This effect is particularly noticeable when a motion-picture camera is panned, and it is one reason for the fact that such cameras usually cover a rather narrow angular field.

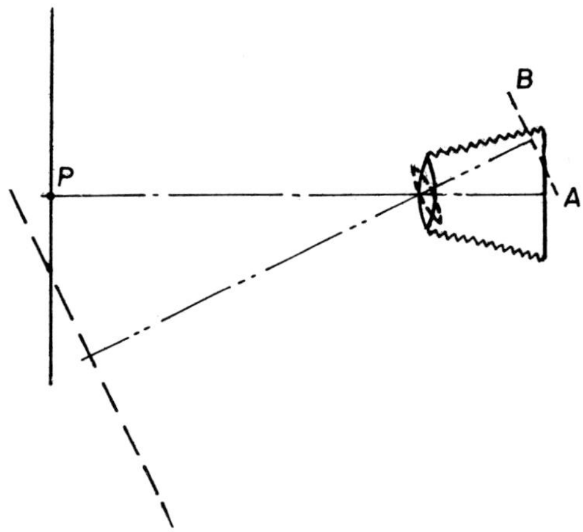

Figure 1.16. Panning distortion. The object *P* lies in the focused plane of the camera in position *A*, but it falls closer than the focused plane when the camera has been rotated to position *B*, giving a larger image.

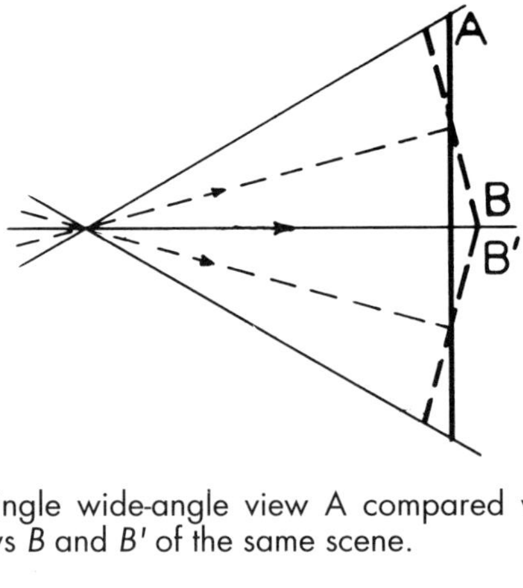

Figure 1.17. A single wide-angle view A compared with two adjacent narrow-angle views *B* and *B′* of the same scene.

When taking a series of photographs intended to be joined end-to-end to provide a panoramic effect, it is essential to ensure that the prints are cut so that the angle out from the axis in one picture is the same as the angle out from the axis in the adjacent picture (Fig. 1.17). One cannot join the middle of one picture to the side of the next, as the images will not match perfectly.

Scanning-Type Slit Cameras

The Panoramic Camera

This is a special type of camera used for photographing broad groups or landscapes, in which the film passes behind a slit in the focal plane of the lens. Two well-known types of panoramic camera exist. In the first type the entire camera is made to rotate slowly about a vertical axis, the picture being taken through a narrow vertical slit on the lens axis close to the film (Fig. 1.18). The film is automatically rolled back past the slit by gearing at such a rate that the point on the film receiving the image is moving at the same speed as the image itself. This type of camera is typified by the Cirkut cameras of 1904, many of which are still in use. Indeed, with such a camera it is possible to repeat a scene many times along one strip of film by letting the camera rotate several times.

In the other type of panoramic camera the lens is made to rotate, during exposure, about a vertical axis through its nodal point, and a shield is attached to the rear of the lens terminating in a slit close to the film

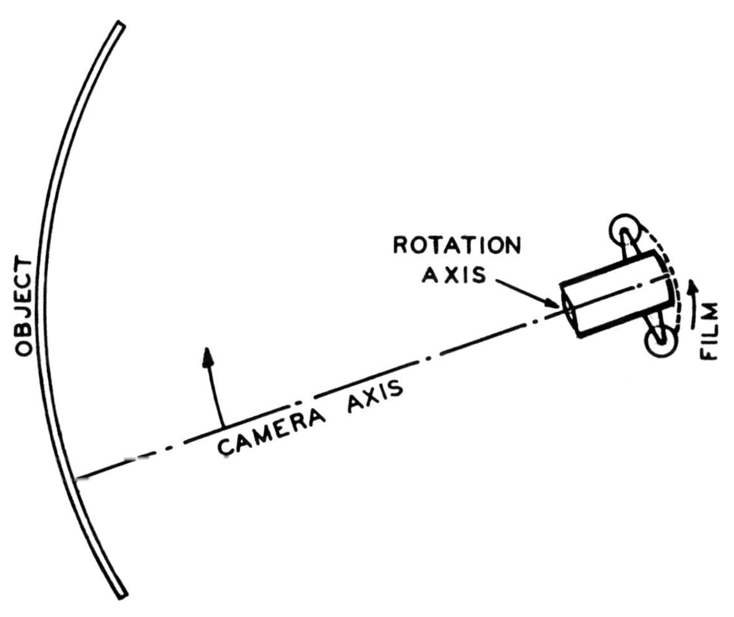

Figure 1.18. The image formed in a Cirkut camera.

(Fig. 1.19). The film itself is held in a cylindrical housing centered about the axis of rotation of the lens, so that during exposure the slit would sweep across the film to form the image. This type of camera was typified by the Al Vista and Kodak Panoram cameras made during the early years of this century. The similar Panon and Widelux cameras are available today.

A curious perspective distortion results from the use of a panoramic camera, since a concentric arc-shaped group of people will appear to be flat on the film. A flat building behind the group will appear to the observer to be cylindrical and convex, the ends of the building being drastically reduced in size.

The reason for this distortion can be seen from the diagram of the camera shown in Fig. 1.19. If a normal fixed lens were used, the image would lie in the flat plane AO. The image in the panoramic camera lies on the cylindrical surface *AP*. At a point in the image corresponding to the field-angle ϕ , the horizontal image dimension from the midpoint of the field is the arc *AP* instead of the line AO; hence the horizontal magnification is $\phi/\tan \phi$. At this same field point, the vertical magnification is $LP/LO = \cos \phi$, since the image is closer to the lens than it would be in a normal camera. We may apply these two magnification ratios to the case of the rectangular object shown in Fig. 1.19(b), giving the image in Fig. 1.19(c).

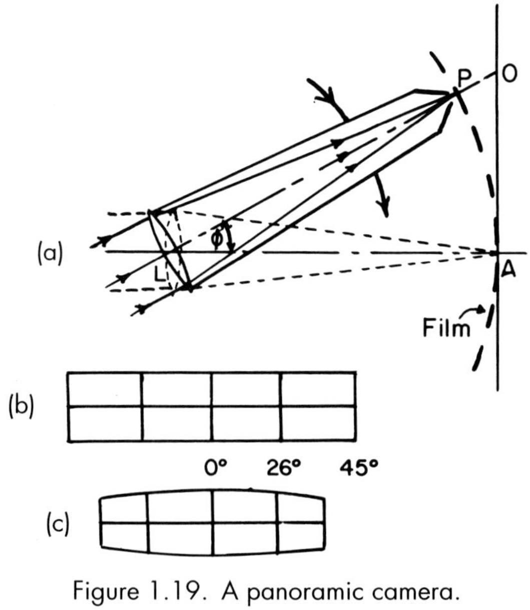

Figure 1.19. A panoramic camera.

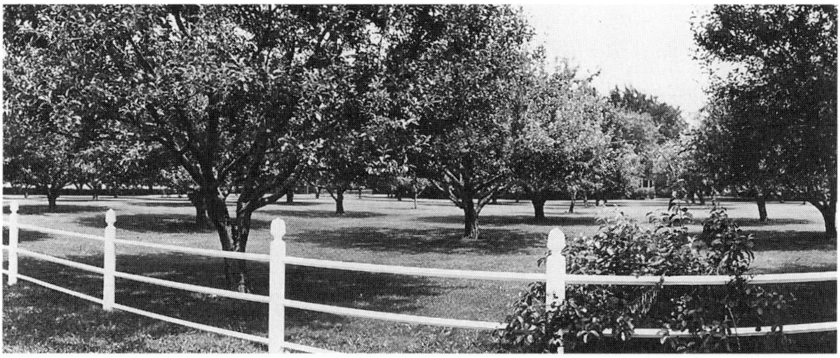

Figure 1.20. Photograph taken with a panoramic camera.

Figure 1.20 shows a photograph taken with a panoramic camera, the fence being actually straight in the original scene.

The Race-Track Camera

This is a special type of camera used to photograph the horses as they cross the finish line in a race. The axis of the camera is pointed along the finish line from one side of the track, the camera being set high so that the near horses will not obstruct those on the far side of the track (Fig. 1.21). A

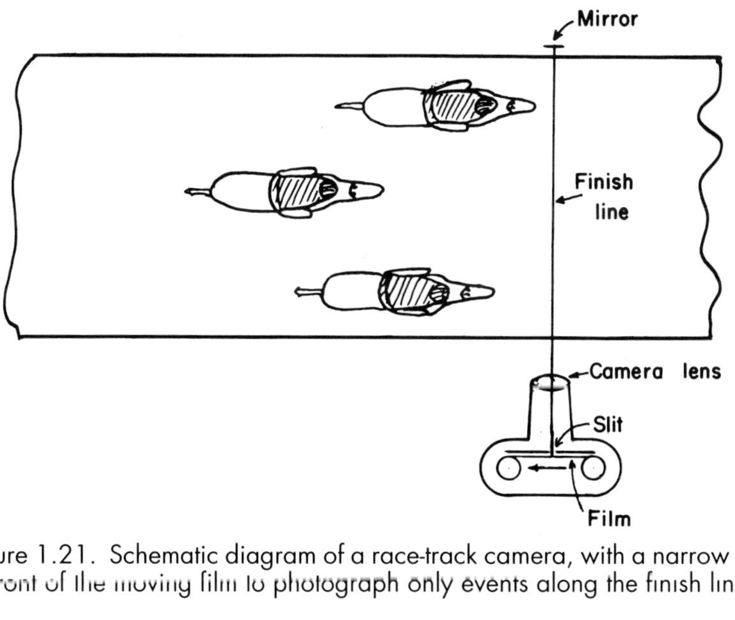

Figure 1.21. Schematic diagram of a race-track camera, with a narrow slit in front of the moving film to photograph only events along the finish line.

vertical slit is mounted immediately in front of the film to limit the imagery to whatever is happening on the finish line and nowhere else. The film is rolled horizontally past the slit from right to left at approximately the same speed as the images of the horses, although of course, the near horses will appear to be moving faster than the distant horses, resulting in some degree of distortion. A large mirror is often mounted on the opposite side of the track to enable a distant horse to be seen clearly even if it happens to be

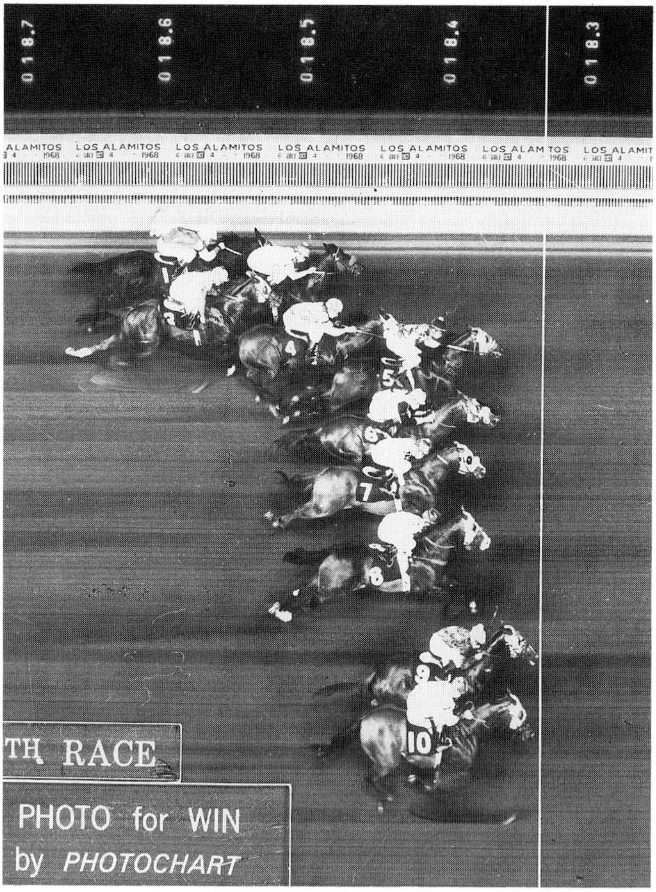

Figure 1.22. An actual photograph of a horse race in which ten quarterhorses were bunched together at the finish line. The time marker at the top was started when the starting gate opened, so that the winning horse's time was 18.33 seconds.

partially obstructed by a closer horse.

It is clear that any vertical line drawn across the picture will represent the finish line at some particular moment of time, and a time scale could be laid along the film indicating the exact instant at which each horse's nose crossed that line (Fig. 1.22). All other objects will appear streaked out horizontally because their images remain stationary while the film is moving.

The Aerial Slit Camera

In recent years great use has been made of a continuously moving film passing behind a fixed slit in the focal plane of an aerial camera. The speed of the film is so adjusted as to be the same as the speed of the image due to the forward motion of the airplane. This may be regarded as a logical extension of the panoramic camera principle. By using two lenses and two slits side by side in front of a broad web of film and setting one slit slightly ahead of the other, it is possible to make a continuous stereoscopic picture of the terrain beneath the airplane (see page 227).

The Periphery Camera

Another application of a slit camera with a moving film is to "unwrap" the periphery of a cylindrical object by mounting it on a rotating turntable, the motion of the film behind a vertical slit being coupled to the movement of the front face of the rotating object. Some amusing pictures have been obtained by mounting a person on the rotating turntable, as shown in Fig. 1.23.

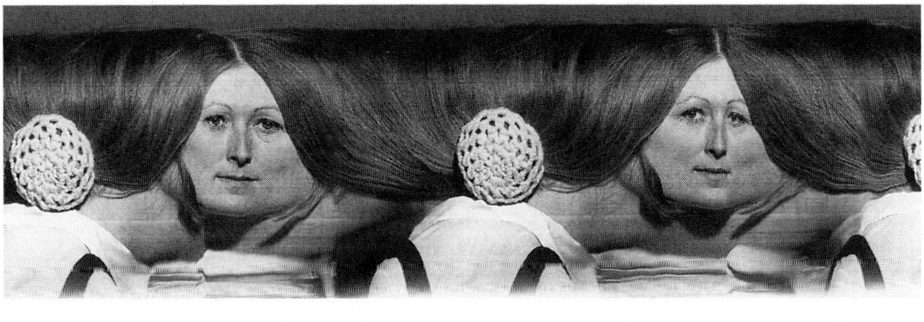

Figure 1.23. A typical periphery camera image. Courtesy A. Davidhazy.

The "Elliptical Wheel" Effect

The elliptical appearance of the car wheel shown in the classical photo

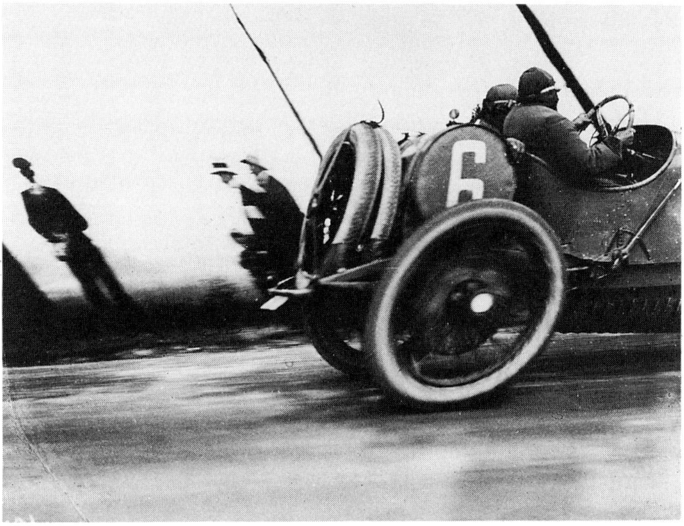

Figure 1.24. Jacques Henri Lartigue, "Grand Prix of the Automobile Club of France, Course at Dieppe" (1911). Gelatin-silverprint 10 × 13$^1/_2$". Collection, The Museum of Modern Art, New York. (Reprinted with permission.)

graph reproduced in Fig. 1.24 is readily explained if it is realized that the picture was evidently taken with a fairly large camera having a focal-plane shutter with a slit moving downward. In this way the ground is imaged first, and other objects above the ground are imaged progressively thereafter. However, because the slit moved relatively slowly, the automobile actually advanced a significant distance during the course of the exposure, leading to the tilted elliptical wheel shown here.

Evidently the photographer tried to keep up with the car by panning his camera, but he failed to do so, and the panning movement caused stationary objects such as the people in the background to be tilted to the left in the photograph. Of course, if the slit in the focal-plane shutter had moved laterally from right to left, as in a Leica camera, the wheel of a car traveling from left to right would appear stretched in a horizontal direction. It is unlikely that this effect would be noticeable in a small 35mm camera, as the shutter curtain moves much faster than the image of a speeding automobile.

Chapter 2

Light Rays and Lens Aberrations

Geometrical Optics

Geometrical optics is the science that deals with the behavior of lenses on the basis of the geometrical relations between the entering and emerging ray paths. It takes no account whatever of the fact that a beam of light is really a train of waves having the power to interfere with the waves in other light beams and the ability to be "polarized." These aspects of light are generally grouped under the heading of physical optics, and will be discussed in the next chapter.

Refraction

For the present, then, we shall regard a ray of light as nothing more than a path or track along which the light energy flows. When a ray strikes the surface of separation between two transparent media, such as air and glass, it is abruptly bent and sets off in the second medium in a new direction. This property is known as *refraction*, the amount of the bending being given by the well-known Snell's law:

$$n \sin I = n' \sin I' \ .$$

Here, n and n' are the refractive indices of the two transparent media, and I and I' are the angles of incidence of the light ray in those media, respectively (Fig. 2.1). The angle of incidence is always measured between the ray itself and the direction of the normal, or perpendicular, to the surface at the point of incidence.

We conclude from this equation that when a ray of light enters a more dense medium, as for example when going from air into glass, it is bent toward the normal, and when going out into air again, it is then bent away

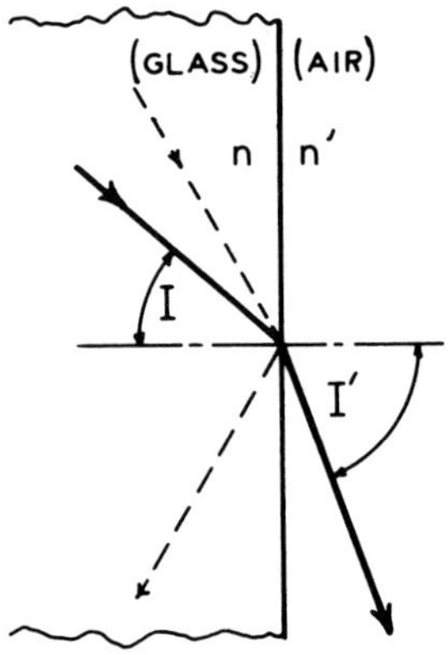

Figure 2.1. Illustrates refraction (solid lines) and total internal reflection (dashed lines).

from the normal. These two effects are indicated in Fig. 2.1. Furthermore, if the angle of incidence inside the glass at the second surface is so large that the sine of the emerging angle of incidence turns out to be greater than unity, then the light ray never does emerge but instead is *totally internally reflected* at the surface and returns into the glass (dotted rays in Fig. 2.1). This fact is commonly made use of in reflecting prisms of all kinds. A typical right-angled reflecting prism of the type used on Photostat or process cameras is shown in Fig. 2.2. If the angular width of the light beam is so narrow that every ray is totally internally reflected, then it is actually better not to silver or aluminize the reflecting face of a prism. The glass-air internal reflection is total (100%), while the best layer of chemically deposited or evaporated metal reflects only about 96% of the light incident upon it. However, camera prisms should be silvered because the angular field is generally quite wide, and the angle of incidence is likely to be less than the critical angle.

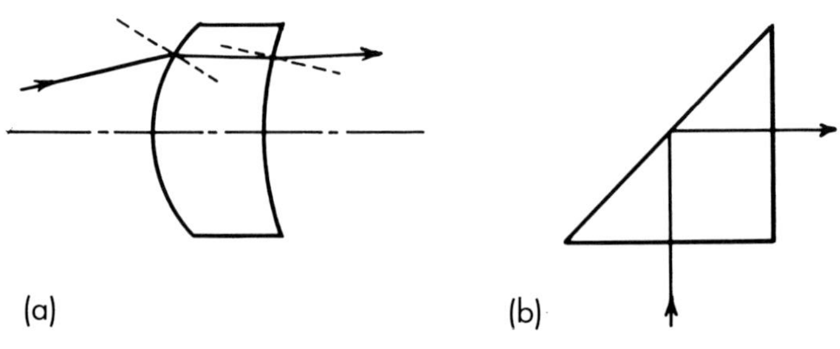

Figure 2.2. (a) Path of a ray through a lens and (b) a right-angled prism.

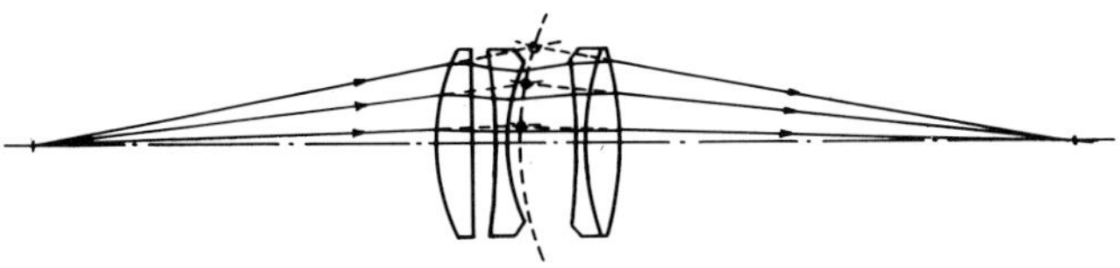

Figure 2.3. The equivalent refracting surface of a lens.

The Focal Length of a Lens

In any ray diagram such as that shown in Fig. 2.14, it will be noticed that over a small area around the middle of the lens all rays come to a common focus. This area of a lens, known as the *paraxial region*, is of special interest as it is free from aberration.

We must now analyze the passage of a bundle of rays proceeding from a point on the lens axis to the corresponding point in the image (see Fig. 2.3). The lens system shown here acts as if it had bent each ray at a certain point inside the lens. The locus of all these points marks out a surface within the lens at which the lens effectively acts, insofar as this particular ray-bundle is concerned. The special cases in which the object is very distant, either to the left or right of the lens, are of particular interest, and in those cases the paraxial portions of the two equivalent refracting surfaces in the lens are called the *principal planes* (Fig. 2.4). The axial point of the principal plane is called the *principal point*.

Because the lens acts as if it were refracting all paraxial rays from a distant object at the principal plane, and as these rays all meet at a single point after refraction, we refer to the distance from the principal plane to the focal point as the *focal length* of the lens.

Since parallel rays can enter the lens from either end, there are obviously two focal lengths for any lens, the *anterior* focus F_1 being at the point where parallel rays entering the lens from right to left cross the axis

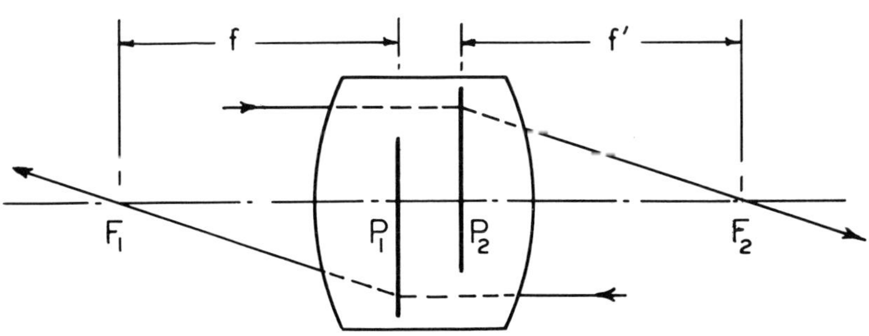

Figure 2.4. The two principal planes and focal lengths of a lens.

after refraction, and the *posterior* focus F_2 being at the point of intersection of parallel rays entering the lens from left to right. The posterior focal length is the one most commonly referred to as the focal length. In the usual case where the focus lies to the right of the associated principal plane, the focal length is said to be positive.

It can readily be proved mathematically that the ratio of the two focal lengths f, f' of a lens is equal to the ratio of the refractive indices of the object and image spaces, or

$$\frac{f}{n} = \frac{f'}{n'} \quad \text{or} \quad \frac{f}{f'} = \frac{n}{n'} \quad . \tag{2.1}$$

Thus, if a lens is used in air, the two focal lengths are exactly equal and we do not need to specify which focal length is meant. Moreover, if the object space is filled with water (see page 51), the anterior focal length will be 1.33 times the posterior focal length.

A normal positive lens has a real focal length and forms a real image that may be reproduced on a photographic plate or on a screen. A negative lens does not form a real image of distant objects, but only a virtual image, which may be seen by looking into the lens from the other side. However, negative lenses do have a place in photography, for example, as the rear member of a telephoto lens (Fig. 2.24), as an amplifier lens when placed in front of a projection lens, and as an attachment for increasing the focal length of an ordinary camera lens.

The reciprocal of the focal length of a lens is called the *lens power*. The most usual unit of power is the *diopter*, which is equal to the reciprocal of one meter. Thus, if we take the reciprocal of the focal length of a lens expressed in meters, we shall have a measure of the lens power in diopters. A lens of 100mm focal length has a power of 10 diopters, and so on. The power of photographic lenses, however, is seldom given in this way, but spectacle lenses and single attachment lenses for a camera are generally so specified. The reason for this is that if two or more very thin lenses are placed together in close contact, their powers may be directly added, but this does not apply accurately to a lens of considerable axial thickness, such as a photographic objective.

When a lens is used to form an image of a very distant object, the size of the image is directly proportional to the focal length of the lens. The actual relation is that if a distant object subtends a small angle ϕ at the lens, then the size of its image on the film will be equal to $f \tan \phi$.

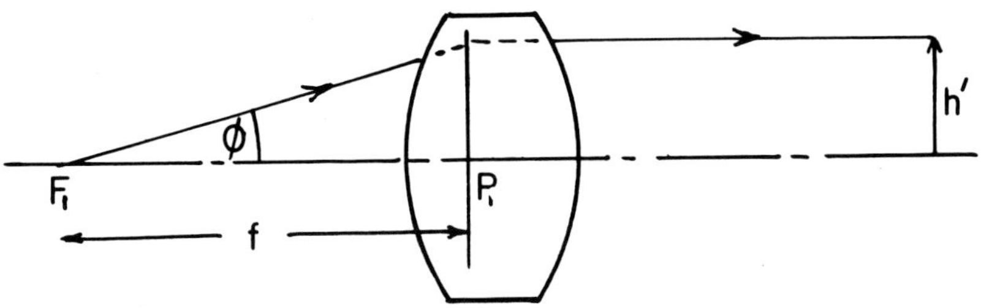

Figure 2.5. A paraxial ray entering a lens through the anterior focus is apparently refracted at the first principal plane and emerges parallel to the axis.

A simple diagram illustrates this relationship (Fig. 2.5). Since the ray passing through the anterior focus F_1 emerges parallel to the axis on the other side of the lens, it is clear that the image height h' will be given by $h' = f\phi$, where ϕ is an extremely small angle expressed in radians. Thus, the anterior focal length f is given by

$$f = \frac{h'}{\phi} \quad \text{or, more precisely,} \quad f = \lim_{\phi \to 0} \left(\frac{h'}{\tan \phi} \right) \tag{2.2}$$

if ϕ is of finite magnitude.

Object-Image Relationships

When a lens is used to form an image of a fairly close object, there are a number of useful relationships connecting the distances of object and image from the lens.

When the conjugate distances are measured from the focal points

If x and x' are the longitudinal distances of object and image from their respective focal points, the image magnification is given by either

$$m = \frac{f}{x} \quad \text{or} \quad m = \frac{x'}{f}. \quad \text{Hence, } xx' = ff' \text{ or } x' = \frac{f^2}{x}. \tag{2.3}$$

This formula is of great value in computing how far the lens in a camera must be moved to focus on an object at a known distance, since x' represents directly the distance between the infinity focus and the image. Thus, if the object is at a distance p from the lens (strictly, from the front principal point), then $x = p - f$, and $x' = f^2/(p - f)$. In some still and small movie cameras, the object distance is measured from the film plane, indicated by a mark ϕ on the top of the camera. In that case, $x' = f^2/(p - 2f)$ approximately.

Front-element focusing

At one time it was common to find the lens in a camera rigidly mounted in its shutter and focused by a movement of the front element only. The advantage of this procedure, apart from the rigid construction, was that in a triplet or Tessar lens the power of the front element was likely to be about three times as great as the power of the entire lens, and consequently the required focusing movement was only one-ninth of the required movement for the entire lens. Front-element focusing has returned to popularity in some zoom lenses, as a movement of the front element has no effect on the zoom mechanism and the image remains in focus during a zoom for any object distance.

When the conjugate distances are measured from the principal points

If p and p' represent the distances of object and image from their particular principal points in a lens, respectively, then it is clear that

$$x = p - f \quad \text{and} \quad x' = p' - f' \ . \qquad (2.4)$$

When these values are substituted in Eq. (2.4), we find that

$$\frac{1}{p'} + \frac{1}{p} = \frac{1}{f} \quad \text{and} \quad m = \frac{p'}{p} \ . \qquad (2.5)$$

These formulae are also of great value, particularly if a lens is very thin and we can assume that the two principal planes are located at the lens itself. However, even if a lens is fairly thick, we can assume with reasonable accuracy that the principal planes divide the lens into three equal parts, and make our measurements from P_1 and P_2 accordingly.

Some handy mental formulas

If a lens is thin, or if we can guess at the position of the principal planes, we can readily construct from Eq. (2.5) the following simple rules that it is well to bear in mind. They refer specifically to the case of a positive lens forming a real image of a real object, all distances and the magnification being assumed to be positive quantities. If virtual images are involved, it is better to return to the original formulas, Eq.(2.5). The relations are

$$p = f\left(1 + \frac{1}{m}\right) \quad \text{and} \quad p' = f(1 + m) \ . \tag{2.6}$$

These formulae may be put into words, as "the image distance is (one plus m) focal lengths, while the object distance is (one plus $1/m$) focal lengths." From this we see that the distance from object to image is given by

$$D = p + p' = f\left(2 + m + \frac{1}{m}\right) = f\,\frac{(1 + m)^2}{m} \ . \tag{2.7}$$

The closest possible distance from the object to the image is $4f$, which occurs at unit magnification when the lens is midway between them. A small longitudinal movement of the lens from its midway position will change only the magnification, and the image will stay substantially in focus. This fact explains a common difficulty experienced by photographers when using an enlarger at or close to unit magnification. The only way to focus an enlarger under these conditions is to raise or lower the negative holder in relation to the easel, or to move the entire enlarger head.

Most camera lenses are designed to operate satisfactorily with an object lying at a distance between infinity and about ten focal lengths; the maximum lens focusing movement is thus about one-tenth of the focal length, representing an extreme image magnification of 0.1. If a lens is designed to be used at a magnification higher than this, it is often called a *macro lens*.

However, some long-focus lenses cannot be moved through one-tenth of the focal length for mechanical reasons. For instance, a 300 mm lens ($f = 1$ foot) is usually made to focus from infinity down to about 25 feet, the focusing movement being then equal to $f/25$ or about half an inch. A 2-inch macro lens used at unit magnification would have to be moved through a distance of 2 whole inches, requiring a complex mechanical mounting. It is better to use a small bellows with such a lens.

Longitudinal Magnification

Ordinary magnification, as we have seen, is measured by the ratio of the transverse dimension of an image to the transverse dimension of the corresponding object. However, it is sometimes important to know the *longitudinal magnification*, that is, the ratio of the longitudinal dimension of an image to that of an object. An example of this arises when we try to determine the longitudinal shift of an image that will result from a given movement of the object along the lens axis.

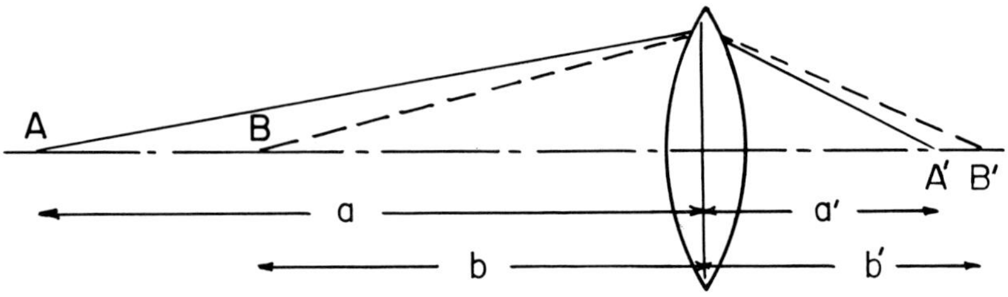

Figure 2.6. Longitudinal magnification.

This situation is illustrated in Fig. 2.6. Here the original object and image are shown at A and A′, at a distance a and a' from the lens. After a longitudinal shift of the object to B, the image will be at B'. The object and image distances now become b and b', respectively.

From Eq. (2.6), we see that

$$a = f\left(1 + \frac{1}{m_1}\right) \qquad a' = f(1 + m_1)$$
$$b = f\left(1 + \frac{1}{m_2}\right) \qquad b' = f(1 + m_2) \ , \tag{2.8}$$

where m_1 and m_2 are the ordinary transverse magnifications when the object is at A and B, respectively.

The *longitudinal magnification* is given by

$$\bar{m} = \frac{b' - a'}{a - b} = \frac{f(1 + m_2) - f(1 + m_1)}{f\left(1 + \frac{1}{m_1}\right) - f\left(1 + \frac{1}{m_2}\right)} \ , \text{ or } \bar{m} = m_1 m_2 \ , \tag{2.9}$$

where all quantities are considered positive in the ordinary case of a real object and a real image.

Of course, if the longitudinal displacement of the object is so small that the magnification remains virtually unaltered, then $\bar{m} = m^2$. An important special case occurs when the magnification is unity and the longitudinal magnification is also unity, for then all three dimensions of the image are the same as those of the object. In every other case the longitudinal depth of the image will be drastically different from that of the object, being much less when the image is smaller than the object and much greater if the image is magnified.

It is interesting to note that if the object and image distances x and x' are measured from their respective focal points, we saw in Eq. (2.3) that $m = \frac{f}{x} = \frac{x'}{f}$. Since $m_1 = \frac{f}{x_1}$ and $m_2 = \frac{x_2'}{f}$, we find that $\bar{m} = m_1 m_2 = \frac{x_2'}{x_1}$ and, if the longitudinal shift is very small, $\bar{m} = \frac{x'}{x}$.

The Image of a Circular Arc

It is a simple consequence of the longitudinal magnification formula that a small circular arc bisected by the lens axis (Fig. 2.7) is imaged as a circular arc of the same radius of curvature, no matter what the magnification may be. The argument behind this fact is as follows: For a small arc, $r = y^2/2x$, and in the image, $r' = y'^2/2x'$. Now $y' = my$ and $x' = m^2x$; hence $r' = r$. This property of a lens has been employed in motion-picture printers where the two films are moved over rollers of the same diameter, no matter whether the image is smaller or larger than the object.

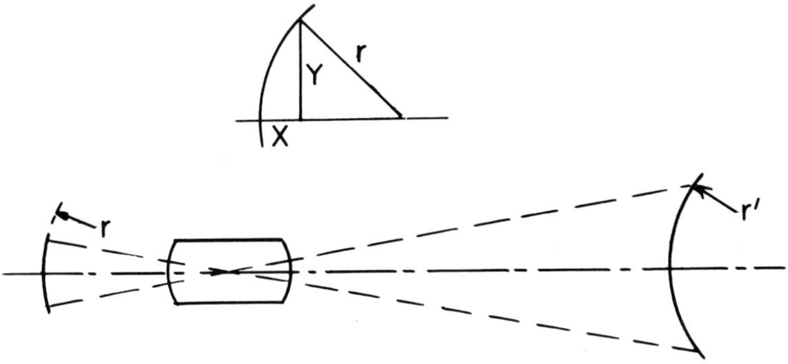

Figure 2.7. The image of a circular arc.

The Image of a Sloping Object

This is an important problem requiring somewhat detailed treatment. In Fig. 2.8, a lens is shown forming a sloping image $A'B'$ of a sloping plane object AB. We must assume that AB is fairly short in order that we can apply the longitudinal magnification formula [Eq. (2.9)]. We now extend the object and image until they cut the middle plane through the lens respectively at K and L. The heights of K and L above the lens axis are given by

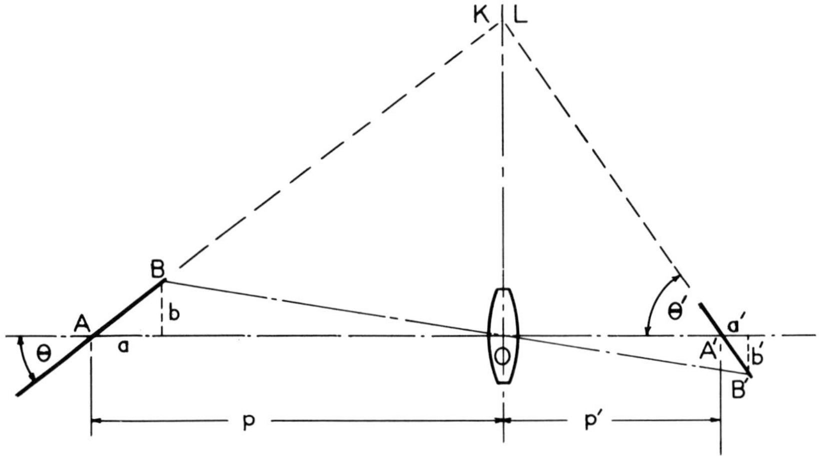

Figure 2.8. The image of a sloping object.

$$OK = p \tan\theta = p\left(\frac{b}{a}\right), \quad \text{and} \quad OL = p'\tan\theta' = p'\left(\frac{b'}{a'}\right). \qquad (2.10)$$

Now if the image magnification at A′ is m, then $p' = mp$, $b' = mb$, and $a' = m^2a$ by the longitudinal magnification rule. Thus, on substituting these values in the formulae for OK and OL, we find that

$$OL = p'\left(\frac{b'}{a'}\right) = mp\left(\frac{mb}{m^2a}\right) = p\left(\frac{b}{a}\right) = OK \; .$$

Hence we reach the important conclusion that *when a lens is used to form an image of a small sloping object, the object plane, the image plane, and the median plane through the lens all meet together at a common point.* This is known as the Scheimpflug condition.

Although we have proved the truth of this rule for small objects only, in actual fact it is valid over surprisingly long distances. Its use in enlargers and projectors for rectifying converging parallel lines is discussed fully in Chapter 9. It has also considerable value in view cameras equipped with "swings" such as that shown in Fig. 1.7. For example, suppose it is desired to photograph a carpet lying on the floor of a room, at the same time showing the vertical lines in the room parallel to each other. The solution is indicated in Fig. 2.9. Here the camera is set with its film plane parallel to the wall AW in order to make the vertical lines parallel, and the lens is then tilted until the film plane CB, the median plane of the lens LB, and the plane of the carpet AB, all meet at the point B. The resulting photograph, Fig. 2.10, shows the carpet sharply defined and all other objects becoming progressively more blurred at increasing heights above the floor. A normal photograph taken with an upright lens is shown in Fig. 2.11.

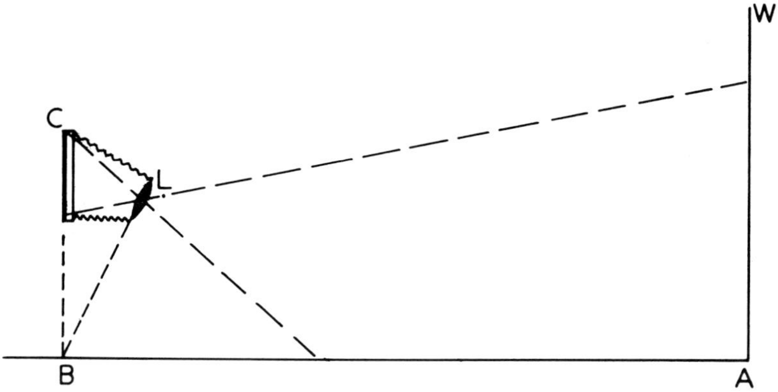

Figure 2.9. Arrangement for photographing a carpet, with vertical lines parallel.

Lens Aberrations

Chromatic Aberration

It is found that all transparent materials are *dispersive*; that is, the refractive index n increases as we go toward the blue end of the spectrum. Thus, the blue part of a ray of white light striking a glass surface will be bent, or refracted, more than the red part of the same ray. One result of this property is that all single lenses possess *chromatic aberration*, to an extent depending on the material of which the lens is made, causing the blue focus to fall closer to the lens than the red focus (Fig. 2.12). A concave (negative) lens behaves in the opposite way, and it was early discovered that a weak negative lens of a highly dispersive material, such as flint glass, could be combined with a stronger positive lens of a less dispersive material, such as crown glass, to form an *achromatic* lens having no chromatic aberration. The negative element, of course, greatly reduces the power of the positive element, so that if we desire to construct a positive achromatic lens of some given power, we must start with a much stronger positive element to which the negative correcting element will be added (Fig. 2.13). Fortunately, there are many different kinds of optical glass available to the lens designer, covering a wide range of both refractive index and dispersion, so that it is not hard to select suitable types of glass to achromatize any lens being designed. Indeed, the only photographic lenses now available that are not achromatized are the simple landscape lenses used in low-cost cameras.

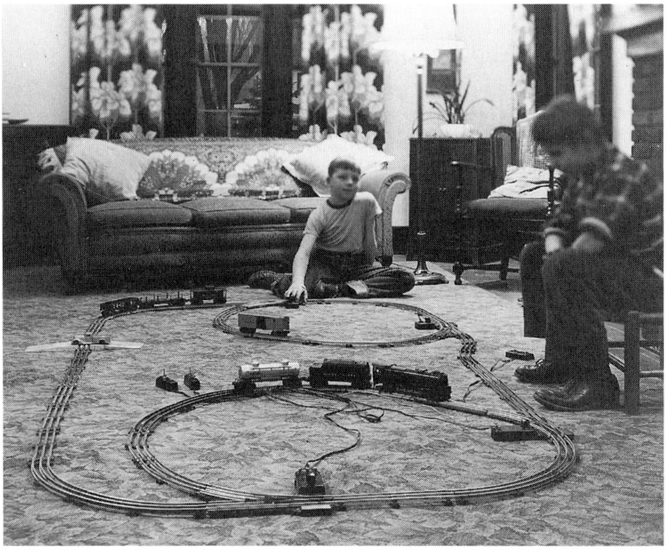

Figure 2.10. Photograph taken with a vertical film and tilted lens, to illustrate the situation in Fig. 2.9. Note that objects on the floor are all in focus, but that objects above the floor become progressively more blurred as they depart from the focused plane.

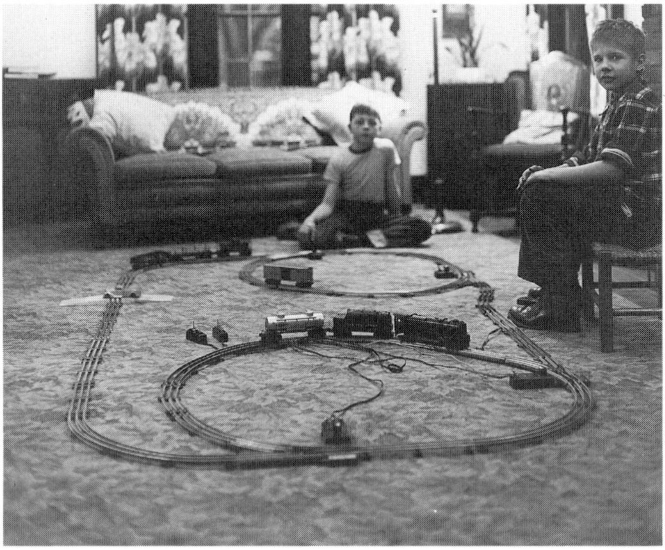

Figure 2.11. The same scene as in the previous figure, but taken with both lens and film vertical. All objects at middle distance are in focus, while those lying closer or further than the focused plane are blurred.

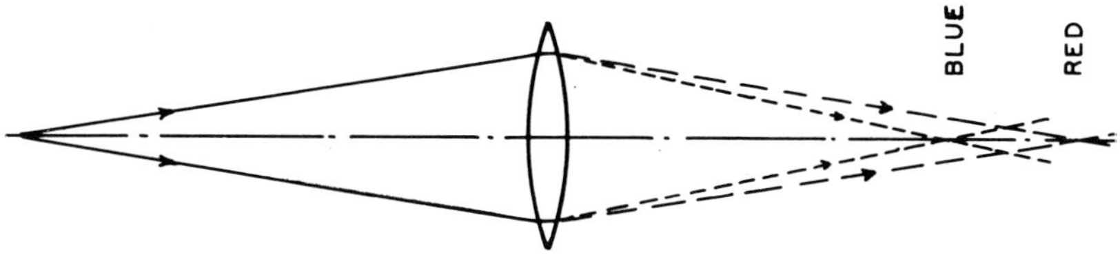

Figure 2.12. The chromatic aberration of a simple lens.

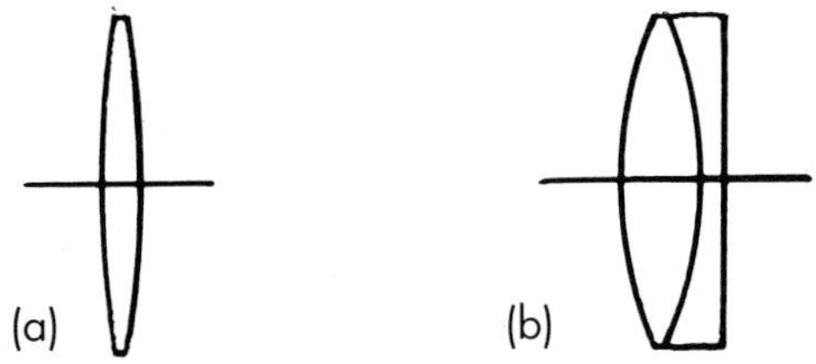

Figure 2.13. (a) A simple lens and (b) an achromatic lens of equal power.

Secondary Spectrum

When we speak of a lens being "achromatized," we mean strictly that the spectrum has been bent back on itself, and that the images in two of the colors of the spectrum have been united at a common focus. The remaining colors, however, will in general depart slightly from the common focus of the two selected colors. Thus, an achromatic lens may show a slight residual of color in the image (*secondary spectrum*), which in a lens of long focus may be great enough to be significant, especially for the ultraviolet and infrared. The light in both these spectral regions comes to a focus slightly beyond the best visual or photographic focus, so that when using infrared-sensitive film, it is advisable to move the lens forward away from the film by about 0.5% of the focal length, to secure the sharpest possible definition.

By the use of crystalline fluorite or unusual types of optical glass, however, it is possible by careful design to construct a lens in which the secondary spectrum is reduced to the point where it is completely negligible. Such lenses, used commonly in three-color process cameras, are known as *apochromats*. The elements are now stronger still, and such lenses are generally limited in aperture to about *f*/10.

Spherical Aberration

When we try to form an image of something by means of a single convex lens, we are likely to be disappointed in the sharpness of the image. This is because the solid cone of rays that enters the lens from a single point in the object does not all come to the same focus on the other side of the lens. Instead, the outer zones of the lens bend (refract) the rays too much, bringing them to a focus somewhat too close to the lens (Fig. 2.14). This phenomenon is called *spherical aberration* since it is due to the general use of spherical surfaces in lenses. The possibility of removing this kind of aberration by the use of aspheric, or nonspherical, lens surfaces continues to prove attractive to both designers and manufacturers. The production of good aspheric surfaces presents a difficult technical problem, but it is slowly being overcome, and today many lenses are being made embodying one or more aspheric surfaces. Of course, an aspheric *plastic* lens is as easy to mold as a lens with only spherical surfaces, but it is difficult to fabricate the mold, and the centering problems become very critical.

Generally, therefore, we correct the spherical aberration of a positive lens by combining with it one or more suitably shaped negative lenses in which the spherical aberration is of equal magnitude but opposite sign. Thus, it becomes a convenient matter for the lens designer to correct both the chromatic and spherical aberration of a positive lens by means of a suitable negative lens element mounted close to it. Sometimes, indeed, as in simple telescope objectives, it is possible to accomplish both these desired corrections by a pair of lens elements having a common inner radius so that they can then be cemented together to make a single unit. Several such units are often employed in a photographic objective, because there the designer must correct other aberrations in addition to the spherical and chromatic, as will be explained later in this chapter.

Zonal Aberration

As in the case of chromatic correction, the combination of a convex and a concave element does not completely remove the spherical aberration of a lens, but it accomplishes this for one zone of the lens only, the other zones

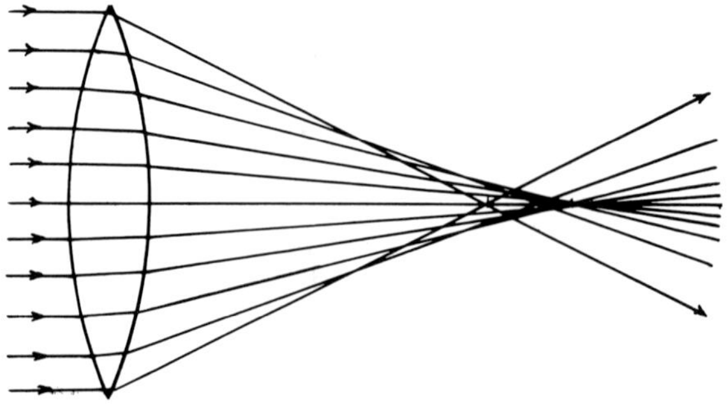

Figure 2.14. The typical ray distribution caused by spherical aberration.

having a small residual of spherical aberration that cannot be removed in any simple fashion. In a good lens, however, this residual is brought to such a small value that it has no noticeable effect on the definition, but in some lenses the zonal residual is sufficiently large to cause the image to shift longitudinally as the lens iris is stopped down. If this phenomenon is observed in a large commercial lens, for example, then it is necessary to focus the image on the ground glass at whatever stop will be used to take the final photograph.

In Fig. 2.15 are shown the graphs of spherical aberration versus

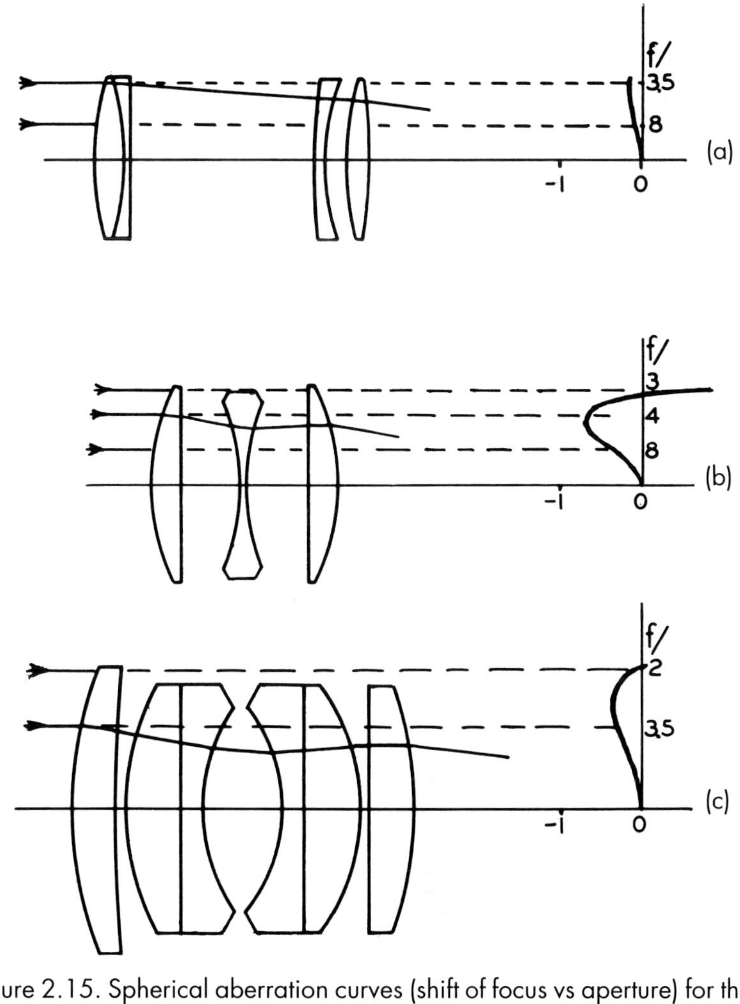

Figure 2.15. Spherical aberration curves (shift of focus vs aperture) for three typical photographic objectives.

aperture for three typical lens systems. Each graph serves to connect the height of incidence of a ray as it enters the lens, with the position of the point at which that ray crosses the lens axis on emergence. Thus, in the middle example (b), the ray entering at *f*/3 (the rim of the lens) will fall slightly beyond the paraxial focus, whereas an intermediate zonal ray at *f*/4 will fall decidedly short, causing some residual zonal aberration. The zonal residual is found to be exceptionally small for the *f*/3.5 Petzval portrait lens (a), and this type of construction is therefore commonly adopted wherever excellent central definition is required. The abscissa scales on these aberration graphs represent percentages of the focal length.

The Oblique Aberrations

We have considered so far the spherical and chromatic aberrations that are found on the lens axis and that thus occur over the entire lens field. In addition to these, there are five *oblique* aberrations that do not appear in the center of the field but that increase progressively at increasing transverse distances from the axis. These aberrations were first studied in detail by L. von Seidel over a century ago, who identified them and gave them the names coma, astigmatism, field curvature, distortion, and lateral color. Although good lenses seldom possess any of these aberrations to a noticeable extent, the photographer should be aware of them and how they may be recognized.

Coma

If a lens exhibits coma, the image of an extra-axial object point is not a single point, but a fan-shaped pattern having a point at one end and a

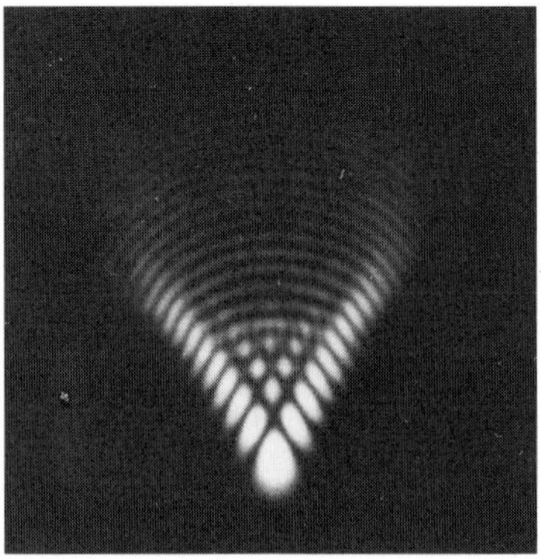

Figure 2.16. A typical comatic star image.

flaring tail at the other. The word "coma" is related to "comet," and indeed the typical coma image shown in Fig. 2.16 resembles some of the comets that have been occasionally observed in the sky. The cause of this unsymmetrical elementary image is that the marginal rays of a comatic lens have a different focal length than the rays passing through the center of the lens, and as the focal length determines the height of the image above the axis, it is clear that when coma is present, the separate zonal images will fall one above another in the image plane (Fig. 2.17).

The focal length of a zone is defined as the distance measured along the ray, from the equivalent refracting surface (principal plane) to the point at which the ray crosses the axis (Fig. 2.18). Hence, if a lens is to be free from both spherical aberration and coma, the second principal plane must in reality be a portion of a sphere centered about the second focal point of the

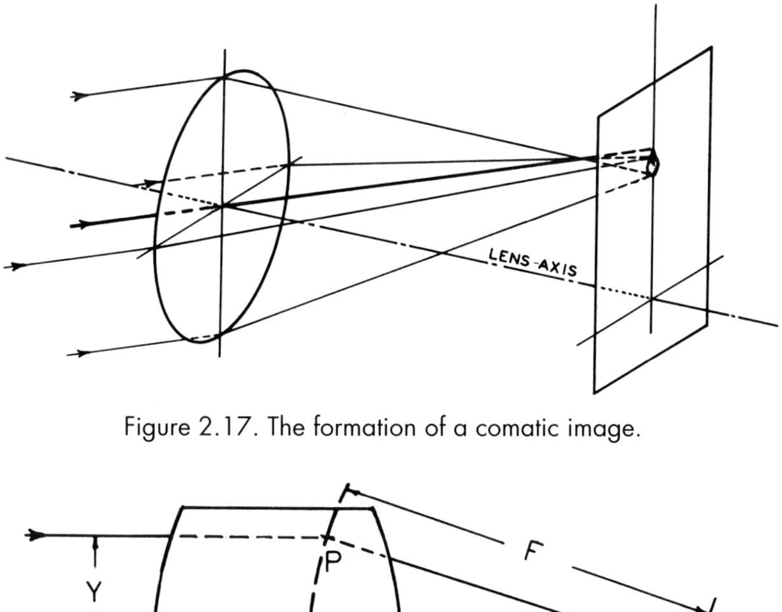

Figure 2.17. The formation of a comatic image.

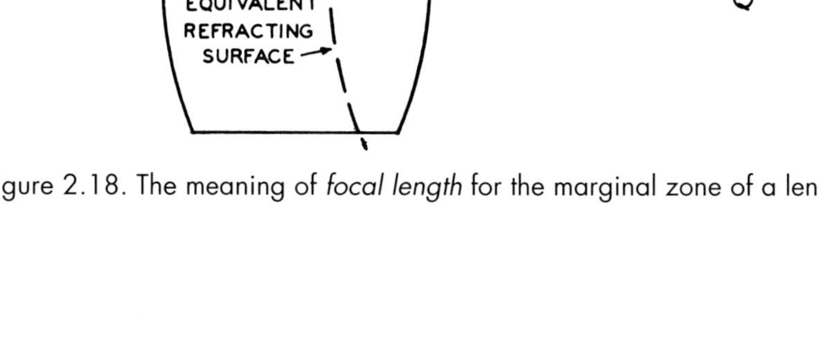

Figure 2.18. The meaning of *focal length* for the marginal zone of a lens.

lens. A lens possessing this desirable type of correction is said to be *aplanatic*. It is thus quite a mistake to imagine that in a "perfect" lens the principal planes are flat and of indefinite extent; in actual fact, the best possible shape for each principal plane is a sphere centered about the object or image point, respectively.

Coma has the effect of producing a rather unpleasant one-sided radial blurring of images lying in the outer parts of the field. Fortunately, it is relatively easy to eliminate in the course of designing a lens, and very few lenses exhibit enough coma for it to be a serious problem.

Astigmatism

Astigmatism is another aberration existing only in the outer parts of the lens field. When astigmatism is present, the emerging cone of rays from an extra-axial object point does not converge to a single point, but instead forms two short focal lines, one lying radial to the lens field and the other tangential to it, both focal lines being perpendicular to the direction in which the light is traveling (Fig. 2.19). The distribution of astigmatism over the field of a lens is commonly represented by two curves (Fig. 2.20), the dotted curve containing the tangential focal lines and the solid curve containing the radial focal lines. The horizontal scale represents the distance of the two focal lines from the image plane as a percentage of the focal length of the lens. It will be evident from this diagram how well the astigmatism of modern anastigmatic lenses can be controlled by careful design.

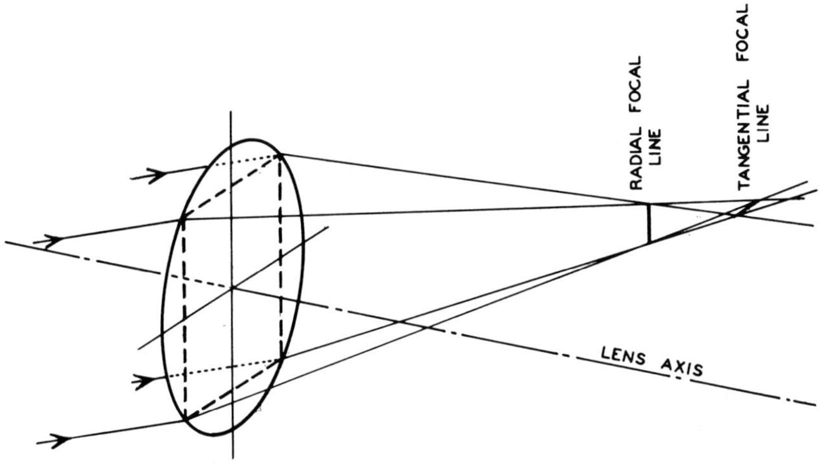

Figure 2.19. The formation of an astigmatic image.

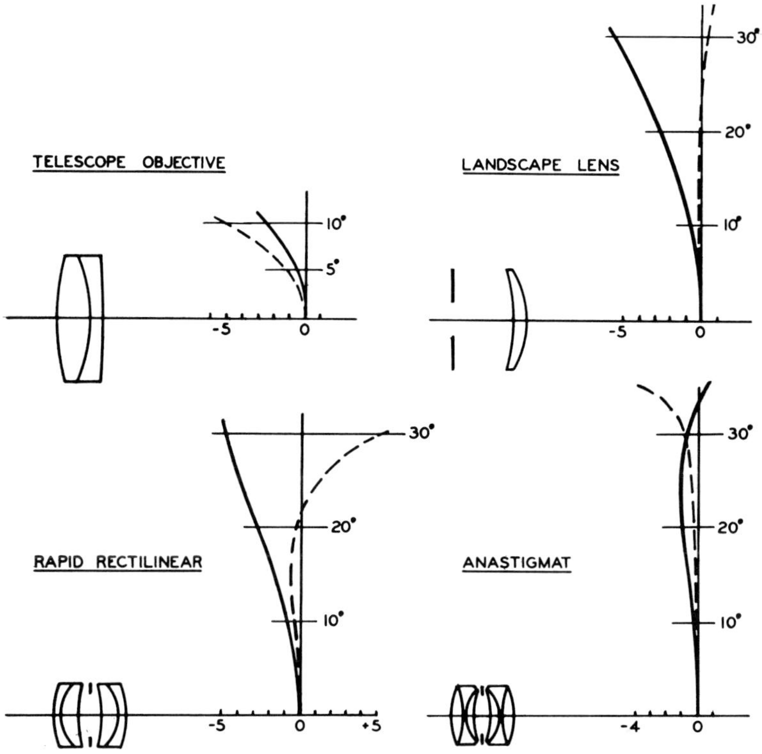

Figure 2.20. Some typical astigmatic field curves.

Field Curvature

A lens that has been corrected for astigmatism is highly desirable, of course, but it will not be satisfactory unless it has also a flat field. As a matter of fact, from elementary considerations, we should expect a simple positive lens to have a curved field, the reason for this being indicated in Fig. 2.21. In that diagram, the axial object-point A is closer to the lens than the oblique object-point B, and we should expect to find that the image of A, at A′, is further from the lens than the image of B at B′, leading automatically to a curved field. If the object is now moved closer to the lens, the discrepancy between the distances of on-axis and off-axis objects becomes greater, and we should expect to find the field becoming more strongly inward-curving. This agrees very well with practice, for lenses such as enlarging lenses that are used at a variety of object distances often show a tendency toward a more inward curving field with near objects, and a more backward-curving field with distant objects.

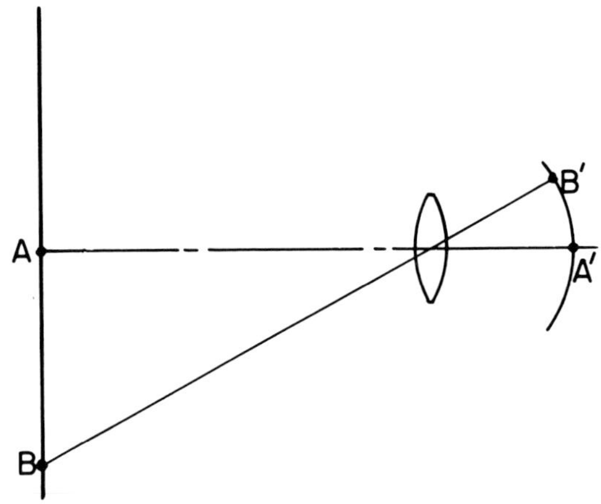

Figure 2.21. The cause of the curved field of a thin lens.

In the 1940s some low-cost cameras were equipped with a simple lens having the usual curved field, which was partially offset by the use of a cylindrically curved film platen. Of course, a cylinder is a poor fit for a spherically curved image surface, but by careful design a compromise was established by which there was some film curvature, enough to fit both the curved field and the flat vertical film section reasonably well.

An interesting example of this arrangement is found in the recent Kodak "Stretch" camera. Here the image is confined to a long central strip of the 35mm film, which is made to fit the basic lens field exactly, with no overcorrected astigmatism. The narrow flat vertical section of the film is not seriously out-of-focus, as it is narrow anyway, and in any case the upper part of the picture is likely to be sky, and the close foreground in the lower part tends to yield a flat image. The long central portion of the negative is greatly enlarged in printing, making a stretched picture eminently suitable for landscapes.

Since 1840 opticians have been striving to overcome the apparently inherent property of lenses by which the astigmatism could be corrected only by introducing field curvature, and vice versa. However, by the 1890s, means were found for overcoming this difficulty, and from that time lenses have been made in which the astigmatism is zero at one obliquity, and the field is substantially flat. Lenses showing this highly desirable property are known as *anastigmats*.

Lateral Color

This aberration represents a change in the focal length of the lens with color, so that the *size* of the image in one color is greater than in another color. The effect if present vanishes in the center of the field, and becomes progressively worse as the obliquity is increased. Lateral color was not a very serious aberration in the precolor days of photography, but even a small amount of lateral color can be very unpleasant in a color photograph, leading to colored fringes along the image boundaries. A particularly sensitive picture is one showing black tree-trunks against snow, for example. Lateral color shares with distortion the property of not being affected by closing down the lens iris.

Because of the importance of good lateral color correction in enlarger lenses that are to be used to make color separations, a convenient laboratory test for it is most desirable. A sheet of blackened glass is taken, such as a fogged and developed photographic plate, and a fine line is scratched through the film tangential to the lens field. Pieces of red, green, and blue filter are mounted behind the scratch adjacent to each other, and an image of the scratch is projected by the lens under test. It is best to photograph the image on panchromatic material so that the image can be studied in detail. Any lateral color that may be present will be revealed by an abrupt step in the line at the filter boundaries, and if this step is large enough to be readily visible under the normal conditions that will be used in observing the separations, the lens should be rejected for color work.

Distortion

The fourth of the oblique aberrations, known as *distortion*, is characterized by a variation of the image magnification over the field of the lens. If the magnification in the outer parts of the field is too great, the corners of a square object will be stretched out disproportionately more than the sides of the square, making the pincushion-shaped figure shown diagrammatically in Fig. 2.22. On the other hand, if the distortion is negative, the corners of the figure will be relatively too small, leading to the barrel-shaped figure shown in this diagram. Distortion of this type does not in any way affect the sharpness of definition, but only the shape of the image. It is readily detected in ordinary photographs, particularly if a vertical line such as the edge of a building happens to lie close to the straight side of the print (Fig. 2.23). Stopping down the lens iris will not affect distortion in any way.

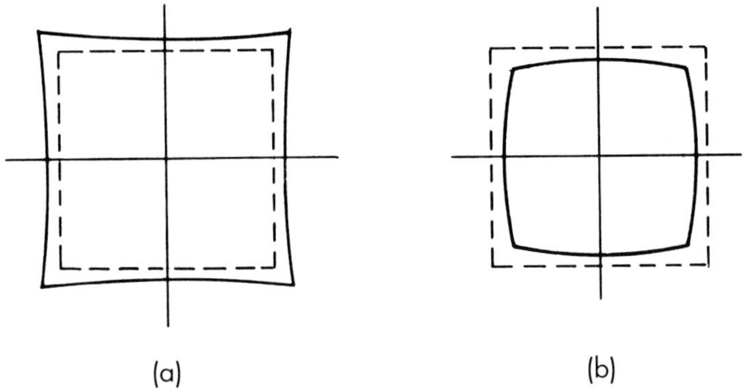

Figure 2.22. Distortion: (a) positive, or pinchusion and (b) negative, or barrel.

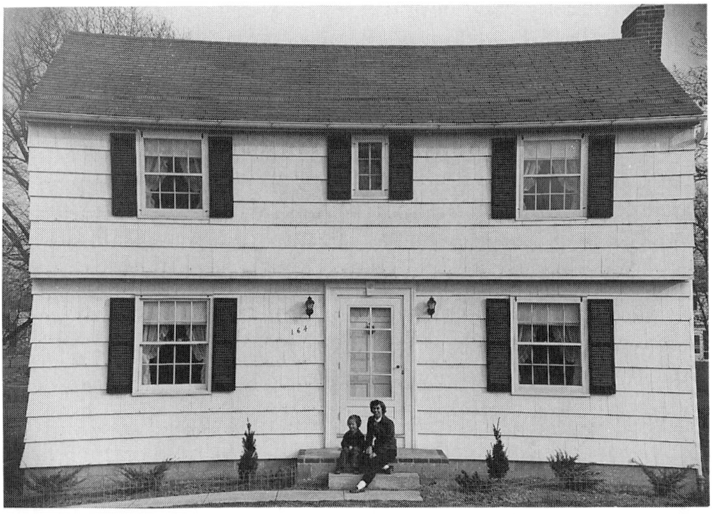

Figure 2.23. A photograph taken with a poorly designed telephoto lens, showing pincushion distortion.

Some Miscellaneous Optical Systems

Symmetrical Systems

A symmetrical lens is one in which every dimension on one side of the diaphragm is repeated on the other side of it, so that the two halves are identical and mounted at the same distance from the diaphragm.

A symmetrical lens has several interesting properties, the chief being the automatic correction of the three transverse aberrations, coma, distortion, and lateral color, when the lens is used at unit magnification. The design problem is thus greatly simplified if a lens is made symmetrical, for each half needs to be corrected only for the longitudinal aberrations, spherical, chromatic, astigmatism, and field curvature. Even when the conjugate distances of object and image from the lens are decidedly unequal, a symmetrical lens is still almost perfectly free from these transverse aberrations. This was one reason for the great popularity of symmetrical anastigmats between about 1890 and 1910. In a "convertible" lens, each half is so well corrected that it can be used alone if desired.

Process and copying lenses are generally of a symmetrical type of construction because neither distortion nor lateral color can be tolerated in such a lens. Distortionless wide-angle lenses for aerial survey work are generally almost symmetrical for the same reason.

The Telephoto Lens

This is a lens of a special type of construction, comprising a positive front component widely separated from a negative rear component. As is indicated in Fig. 2.24(a), this form of construction has the property that the second principal plane lies outside the positive end of the system. The total length from the front lens vertex to the image plane is therefore less than the focal length, a real advantage in lenses of very long focal length. However, it is more difficult to achieve good aberration correction in a telephoto lens than in a lens of a more normal type; consequently, a telephoto lens is used primarily when compactness demands it.

It is common practice with some manufacturers to call any lens a "telephoto" if its focal length is longer than the normal value for that

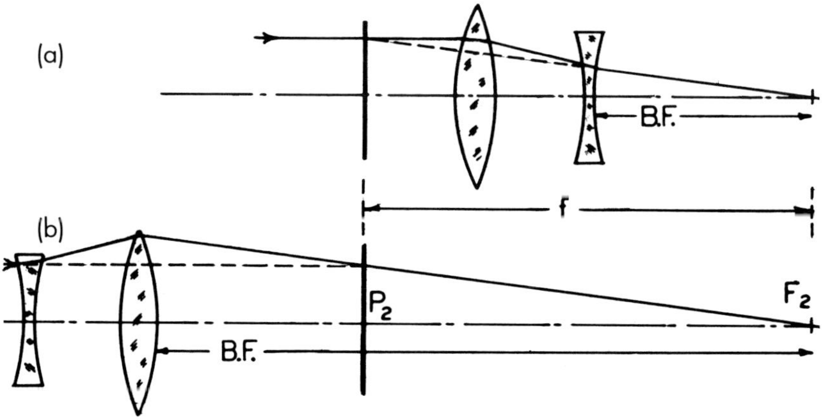

Figure 2.24. (a) The telephoto and (b) the reversed telephoto lens system.

particular type of camera. This practice is, however, misleading and should be avoided. Telephoto lenses are discussed on page 148.

The Reversed Telephoto Lens

If the negative component of a telephoto lens is turned toward the distant object, a system is obtained in which the *back focus*, or distance from the rear lens vertex to the film plane, is exceptionally long, often as long as or longer than the focal length of the lens [Fig. 2.24(b)]. Such a system is useful whenever a long clearance is required between lens and film. The earliest application was in the 3-strip Technicolor camera where a beam-splitting prism was mounted behind the lens to separate the three colored images. Then it became common in short-focus lenses for small movie cameras to provide space for a reflex viewfinder or a turret and the shutter. Today, reversed telephoto lenses are universally employed in wide-angle lenses for single-lens reflex (SLR) cameras, where the tilting mirror and focal-plane shutter require at least 35 mm clearance.

The Simple Magnifier or "Loupe"

When a lens is held close to the eye to act as a magnifier, the object must be placed approximately at the anterior focus of the lens in order that each point in the object will send a beam of parallel light to the eye. The apparent angular size of the image, as seen by the eye, will then be equal to the angle subtended by the object at the front principal point of the lens. The entire advantage of using such a magnifier is, therefore, that it enables us to bring an object exceptionally close to our eyes without having it go out of focus.

Assuming that the least distance of distinct vision for an ordinary unaided eye is 10 inches or 250 mm, the *magnifying power* of a lens of focal length f, when held close to the eye, will thus be equal to the simple ratio

$$\text{Magnifying Power} = \frac{250\ \text{mm}}{f} \quad .$$

The Compound Microscope

The compound microscope consists of an *objective* lens that forms a real magnified image of the object, and an *eyepiece* to examine this image. If the objective lens magnifies m_1 times at a certain tube length and the eyepiece has a magnifying power of m_2, then the overall magnifying power of the microscope at that tube length is equal to the product of m_1 and m_2. Since these magnifications are marked on the objective and eyepiece, respectively, the user can readily determine their product. The working distance below the objective of a compound microscope is much larger than that of a simple magnifier of the same ultimate magnifying power, but the image is inverted unless special erecting prisms are added below the eyepiece.

The normal tube length of the modern microscopes is about 6 inches from the mounting seat of the objective to the image, and the magnification engraved on the objective will be approximately correct at that distance. Direct photography of small objects at considerable magnification through an objective alone is often referred to as *photomacrography*. For greater magnifications the eyepiece must be added and the film moved back to a point several inches beyond the eyepiece. The overall magnification is then the product of the engraved objective magnification and the eyepiece magnification computed by the relation (image distance/focal length) minus one, i.e., $(p'/f) - 1$. This process is referred to as *photomicrography*, and it should not be confused with the similar term "microphotography," which refers to the taking of very tiny photographs of extended objects such as documents.

A related term is *microfilm*. This refers to a series of documents or illustrations along a strip of motion-picture transparency film, 16 or 35mm, with the usual perforations. The demagnification may be $^1/_{20}$ or $^1/_{30}$ times full size, the film being read in a suitable magnifying reader. If the successive images are formed on a 4×5 inch transparency, the result is generally referred to as a *microfiche*.

A Lens Used under Water

In underwater photography, the camera is usually enclosed in a watertight housing with a plane glass front window (Fig. 2.25). The refraction of light

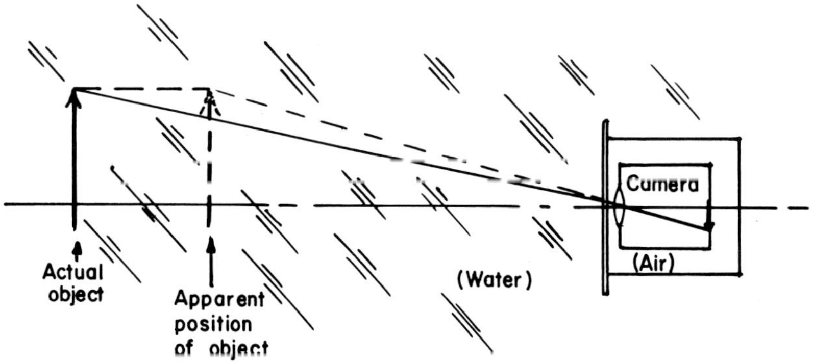

Fig. 2.25. Underwater photography.

at the air-water interface has the effect of bringing the object closer to the camera. The apparent position of the object, as indicated in this diagram, is only three-quarters of the true object distance because the refractive index of water is 4/3; hence, the image of an underwater object is larger than it would be in air, by this ratio.

If we regard the plane front window as being part of the camera lens, we have the situation in which the object and image spaces have different refractive indices, and consequently the two focal lengths of the lens are different. Equation (2.1) indicates that in this case the anterior focal length f is 1.33 times the posterior focal length f'. Equation (2.2) tells us that since $h' = f \tan \phi$, the image height for a distant object subtending an angle ϕ will be 1.33 times as great as if the lens were used in air.

Besides this increase in image size, the refraction of light at the window introduces every kind of aberration, and attempts have been made to offset these by the insertion of one or more lens elements inside the housing in front of the camera. A much better approach is to design the camera lens from the start for this particular application, as in the Nikonos 35mm cameras.

Some workers have tried using a spherical dome in front of the camera with the camera lens situated at its center of curvature. While this arrangement eliminates some aberrations such as distortion, it greatly increases the focal length of the system and introduces many other aberrations, so it is generally felt that a plane window is a better solution.

The Field Lens

Whenever a long optical system is employed, in which a first objective is used to form an internal real image that is in turn reimaged by a second objective, it is necessary to place a field lens at or close to the plane of the intermediate real image, to image the aperture of the first objective upon that of the second objective. In this way, the oblique rays of light that form the outer parts of the internal image will be bent around and directed toward the second objective. If the field lens is omitted, the corners of the image are likely to appear quite dark. The situation is illustrated in Fig. 2.26.

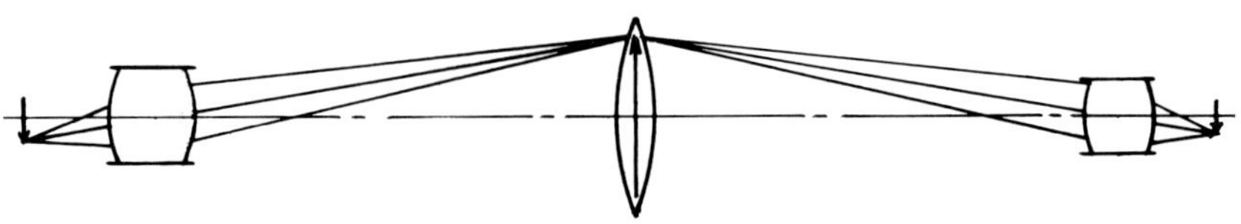

Figure 2.26. The use of a field lens in a TV multiplexer.

Fresnel Lenses

For many purposes where good image formation is unnecessary, such as in condensers, spot lights, and field lenses, it is possible to collapse a lens into a flat plane in small steps, as shown in Fig. 2.27. Here the part *ab* of the original thick lens has been stepped in to *a′b′*, in order to reduce the mass of the lens and to prevent it from becoming too thick. Similarly, the part *bc* is stepped in to *b′c′*, and so on across the whole lens. Lenses of this type were recommended by Fresnel in 1822 for lighthouses, ship's lanterns, etc. More recently, they have become familiar to the photographer in spotlight condensers. In this case the lens is molded in glass without great precision, the steps being bold and few in number.

However, for use as a field lens it is possible to mold a Fresnel lens in a transparent plastic, with a very large number of minute steps so close together that the steps themselves can scarcely be seen without a magnifier. In this form we have a thin sheet lens that can be cut out with scissors and mounted in any convenient way. An excellent application is to place such a lens close to the ground glass of a view camera or a translucent projection screen to increase the brightness of the corners of the picture (see page 135). Once a suitable mold has been made, which is quite costly, the lenses themselves are not nearly as expensive as the familiar ground-and-polished glass lenses. The condenser in an overhead projector is generally of the Fresnel type (see page 219).

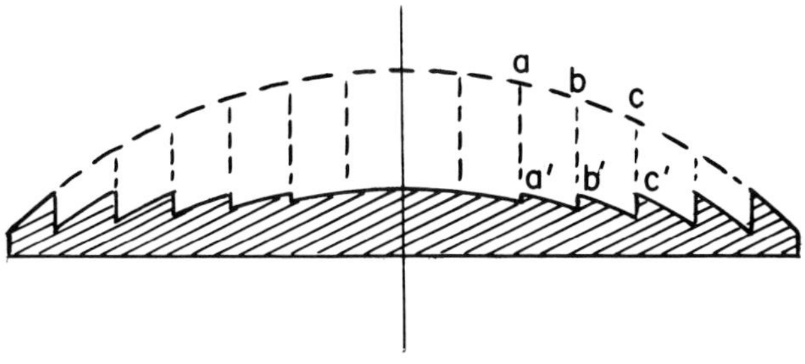

Figure 2.27. The Fresnel lens.

Mirrors

Although curved mirrors are rarely used to form images in photography (the few exceptions are described on page 163), plane mirrors are commonly used to deflect a beam of light into a new direction without introducing any aberrations or other defects in the image. A familiar example is the 45° mirror in a single-lens reflex camera.

However, it must never be forgotten that a plane mirror gives *left-right reversal* to an image, which is the same effect that we get when a transparency photograph is turned around and viewed through the wrong side. Left-hand and right-hand thus become interchanged. This phenomenon is entirely independent of which "way up" the image is seen, and indeed we find that a lens such as that in a camera or a slide projector inverts the image without any left-right reversal. It is important to keep a clear distinction between inversion such as is caused by a lens and the left-right reversal caused by a mirror.

An *odd* number of plane mirrors always gives a left-handed image, but this effect can be counteracted by looking at the picture from the other side. Thus, in a Photostat camera in which the original copy and the final print are *both* looked at from the position of the camera lens, a left-right reversal will occur unless it is properly neutralized by means of a plane mirror or prism in the beam. In a slide projector as ordinarily used, the projectionist stands behind the projector and looks toward the screen. He holds the slide as he wishes it to appear and then rotates *it in its own plane* through 180° before dropping it into the slide carrier. No left-right reversal will occur if this procedure is followed. However, if the audience is on the far side of a translucent screen, then a left-right reversal is required, and the projectionist must turn each slide around a transverse horizontal axis before inserting it into the carrier, to give it the necessary reversal. If a mirror is used, this reversal becomes unnecessary.

A plane mirror also reverses the *direction of rotation* of an image about the light beam as an axis. For example, if a slide is rotated clockwise as seen by the projectionist when standing behind the projector, the image will turn in a clockwise direction on an opaque screen. But the audience sitting beyond a translucent screen will see the image turning in a counterclockwise direction unless this effect is counteracted by a plane mirror.

An important photographic application of mirrors was found in so-called "one-shot color cameras" such as the National Photocolor and Devin Tricolor cameras (Fig. 2.28). In these cameras, two very thin stretched membranes are lightly aluminized to render them partially reflecting so that they will serve as beamsplitters. The first mirror reflects about one-third of the light to the first film holder, and the second mirror divides the remaining light equally into two beams going to the other film holders. Suitable red, green, and blue tricolor filters are inserted in front of the films to give directly the desired color-separation negatives. The lens must have rather a long focal length, and of course it must be exceptionally well color-corrected. The short-lived Canon Pellix camera contained a

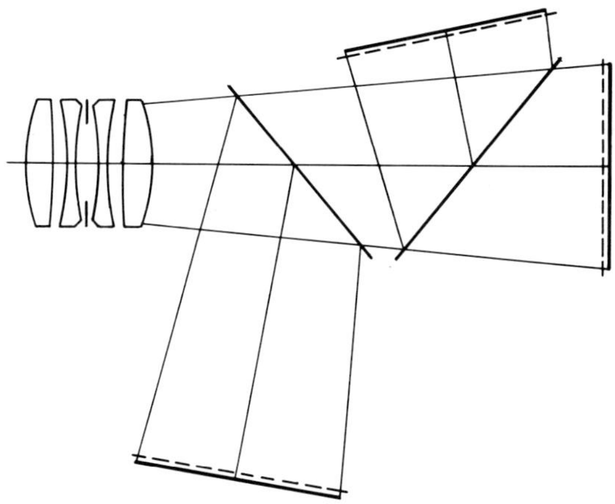

Figure 2.28. The arrangement of pellicle reflectors in a one-shot color camera.

thin plastic pellicle mirror to reflect about 35% of the light to the viewfinder and transmit 65% to the film. By not having to raise a mirror immediately prior to making the exposure, this camera was steadier and quieter than a conventional SLR.

Prisms

For various reasons, chiefly because of simplicity in mounting and cleaning, a plane mirror is often replaced by a right-angle prism (Fig. 2.2). The entering and emerging optical axes are perpendicular to the transmitting faces of the prism, the ray being totally reflected at the hypotenuse face because the angle of incidence is 45°, which is greater than the critical angle (page 28).

Optically, a prism used in this way is equivalent to a plane mirror plus a parallel-sided slab of glass of thickness equal to the base of the prism. The effect of inserting such a thick plate of glass into a converging beam of light is to shift the image longitudinally away from the object by a distance equal to about one-third of the thickness of the plate (Fig. 2.29). The shift is equal to $[(n-1)/n]t$, which for glass having an index close to 1.5, comes out to almost exactly $t/3$.

The Straight-Line Reverser

A prism was generally used in front of the lens of a Photostat camera to give the necessary left-right reversal in the opaque copy. This had the advan-

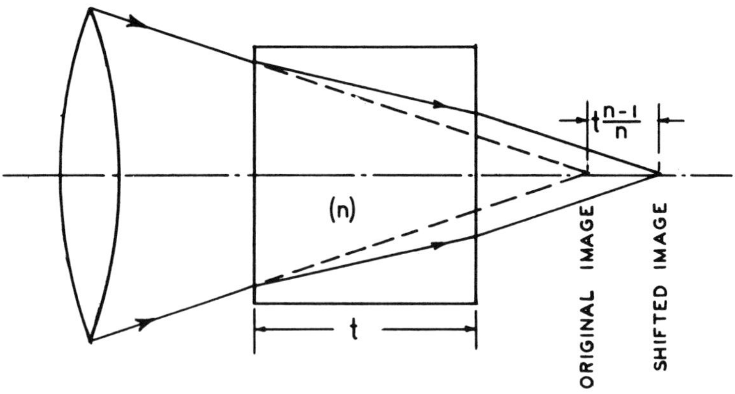

Figure 2.29. The shift of an image caused by inserting a thick parallel plate of glass.

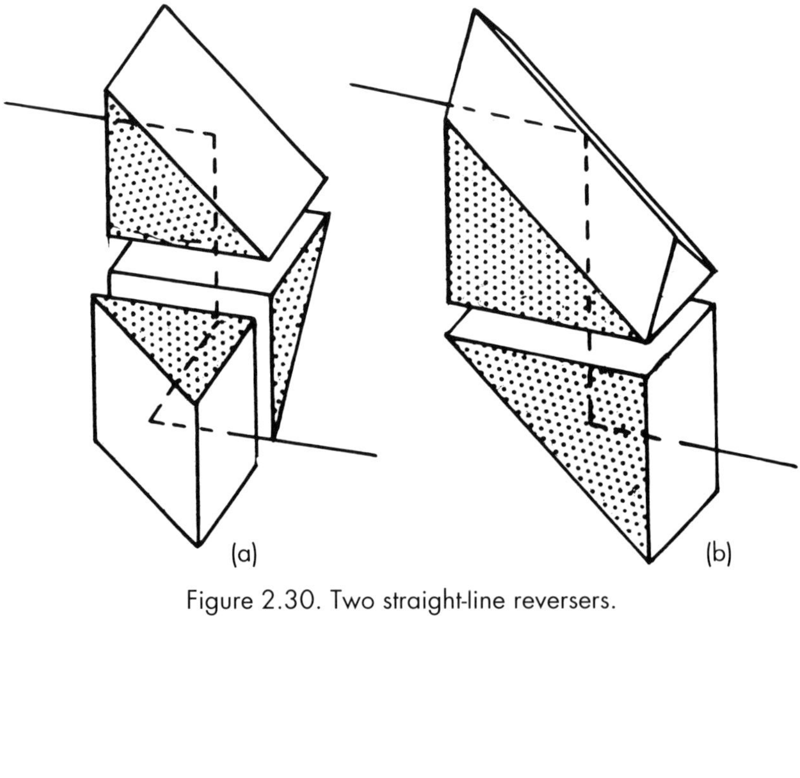

Figure 2.30. Two straight-line reversers.

tage that the line-of-sight was turned through 90°, requiring a horizontal easel. In process cameras a set of three mirrors is often used to give a left-handed reversal to the copy, the mirrors being arranged in such a way that the direction of the optical axis remains unchanged. Such a device is called a *straight-line reverser*. Although actually mirrors are used, the diagrams in Fig. 2.30 show prisms to indicate more clearly how this device is constructed. Another type of straight-line reverser uses a roof prism mounted over a simple right-angle prism, as shown in Fig. 2.30(b).

Chapter 3

Light Waves and How They Behave

Physical Optics

In the previous chapter we considered the properties of a light beam regarded as a group of rays. The paths of these rays through a lens produce its well-known focal properties and aberrations. However, light is actually a form of wave motion, and the study of the wave nature of light constitutes the science of physical optics. Some of the wave properties of light are of particular interest to the photographer, and these will be outlined in the present chapter.

Velocity, Frequency, and Wavelength

Any moving wave train must possess a velocity, a frequency, and a wavelength. In the case of light, these quantities are very extreme compared with their values for some of the other forms of wave motion with which we are familiar, such as sound or water waves. The *velocity* is extremely high, namely, 186,000 miles per second in vacuum and about two-thirds of that velocity in a transparent substance such as glass. The *frequency* of light waves, or the number of waves passing a given point in a second, is also extremely high, being of the order of five hundred million million vibrations a second. The exact figure varies somewhat with the color of the light, from about 4×10^{14} vibrations per second for the red end of the spectrum to over 7×10^{14} vibrations per second in the violet region. The *wavelength* of light, or the distance from crest to crest in the wave train, also varies with color, its value in vacuum being about 0.75 micrometers in the deep red and 0.40 micrometers in the extreme violet. (A micrometer, abbreviated μm, is one-thousandth of a millimeter, or one twenty-five thousandth of an inch.)

We must thank the extreme smallness of the waves of light for the fact that a narrow beam of light travels in a straight line. A train of long waves such as those of sound or radio tends to go around corners, and for them a lens or anything like it becomes almost impossible to construct.

Reflection of Light by Glass

The great French physicist Augustin Jean Fresnel first studied in a quantitative fashion the reflection of light waves at a transparent surface. He found that when light strikes such a surface, a portion of the light is reflected and the remainder is transmitted. The magnitude of the reflected fraction depends on the refractive index n of the glass. This is because refractive index is physically nothing but the ratio of the velocity of light in air to its velocity in glass, and the greater the refractive index, the more the light will be retarded as it enters the glass and the greater the fraction of the light reflected at the surface.

Fresnel's formula for the fraction of light reflected at the surface of a glass of refractive index n, for perpendicular incidence, is

$$R = \frac{(n-1)^2}{(n+1)^2} \quad .$$

The value of R for some typical refractive indices is given in the following table.

Refractive Index, *n*	Perpendicular Reflectivity, *R*
1.33 (water)	2.0%
1.5 (window glass)	4.0
1.6 (flint glass)	5.3
1.7 (dense flint glass)	6.7
1.8 (very dense flint)	8.2
1.9	9.6
2.0	11.1
2.42 (diamond)	17.2

The reflectivity of a surface at oblique incidence is greater than this value.

The reflection of light by the polished glass surfaces of a lens has long been a nuisance to photographers because it leads to ghost images and flare spots, which are fully discussed in a later chapter (see page 124). Fortunately, in recent years means have been found for reducing the reflectance and eliminating the ghost images.

Polarized Light

Natural light consists of a transverse vibration that is perfectly random in its orientation within the beam, but it is possible by various simple means to produce a beam of so-called *polarized* light in which the vibrations all lie in one direction. For example, a beam of light that has been reflected obliquely from a glass plate at an angle of incidence such that its tangent is equal to the refractive index of the glass, is found to be completely polarized. The vibrations in this case lie in a direction perpendicular to the plane of incidence containing the incident and reflected rays.

As it may be desired to produce polarized light by this means, the following table of polarizing angles may be useful.

Refractive Index, *n*	**Polarizing Angle of Incidence**
1.33 (water)	53.2°
1.5 (plate glass)	56.6°
1.6 (flint glass)	58.0°
1.7 (dense flint glass)	59.5°

In attempting to photograph an oil painting behind glass, for example, it is possible to eliminate the reflection of the lamps from the picture glass by placing the lamps in such a position that the reflected light is fully polarized and then extinguishing it by adding a polarizing filter over the camera lens (see page 169).

Other devices for polarizing light are the Nicol prism and the Polaroid filter. Since a polarizer selects only half of the light and rejects the rest, no polarizing device can possibly have a transmittance greater than 50%, and most of them transmit even less because of surface losses or internal absorption.

If two polarizers are used in succession, set so that the plane waves transmitted by the first are also transmitted by the second, a further small reduction in intensity will occur, and the light will remain plane polarized. If the second polarizer is now rotated about the light beam as an axis through an angle θ, the light intensity will be further reduced by a factor of $\cos^2\theta$ until, when $\theta = 90°$, the light is entirely extinguished. A pair of polarizers thus provides a convenient means for varying the intensity of a beam of light at will. Polarizing spectacles have been frequently employed in this way for separating the two images in projection stereoscopy and vectographs (page 235).

Interference

If two beams of light originating from the same source are brought together in such a way that the successive wave crests and troughs in one beam fall upon those in the other beam, *interference* will occur, forming either a standing pattern of light and darkness, or a system of colored rings or fringes. An application of interference that is of importance to photographers is found in the coating of lens surfaces to reduce the amount of light reflected by them (page 129). The light reflected from the outer surface of the coating interferes with the light reflected from the interface between the coating and the glass itself, and if the thickness of the layer is correct, it is possible to extinguish the reflected light entirely in this way, at least for one wavelength; reflected light of other wavelengths will be drastically reduced in intensity.

The "Newton's ring" pattern that is sometimes seen when a film negative is held between glass plates in an enlarger or projector is also a manifestation of interference. In this case, interference occurs between the direct beam of transmitted light and the much weaker beam of light that has been reflected first at the film and then at the glass. The interference fringes produced in this manner are faint because of the great inequality in the brightness of the two interfering beams.

Diffraction

A much more subtle and more troublesome consequence of the wave nature of light is *diffraction*, by which is meant the very slight lateral spreading of a beam of light as it passes an opaque object. In photographic lenses, diffraction appears in two different forms: (a) the "spikes" that are sometimes seen radiating outward from the image of a small bright light against a dark background, and (b) a slight blurring of the image at very low relative apertures. These phenomena will be considered separately, although they are closely related in nature.

Diffraction Spikes in Night Photos

Suppose we have a lens with a five-bladed iris, which is stopped down to a point where the aperture is definitely five-sided instead of being circular. Light waves passing through this opening will be slightly spread out by diffraction in a direction perpendicular to the edge of each diaphragm blade. There will thus appear on the film ten short radial "spikes" projecting from the image of a distant lamp, for instance, when seen against a dark background. This phenomenon is relatively common in night photos (two or three examples may be seen in Fig. 6.15). The cure is either to work with the iris wide open or to use an iris with so many blades that its closed-down shape is virtually a circle.

Bright lights photographed through a window screen at night often show a cross. This is caused by diffraction of light as it passes the horizontal and vertical wires in the screen. In an oblique photograph taken through a screen, the bars of the cross will not be at right angles to each other. In star photographs made with a reflecting telescope, a similar cross is often seen, especially in the images of bright stars. This is caused by diffraction at the support structure for the secondary mirror in the telescope.

Diffraction Blurring at Very Low Lens Apertures

In the preceding section it was stated that when light passes the edge of a diaphragm blade in a lens, it is slightly scattered in a direction perpendicular to that edge. Consequently, if the diaphragm is circular, light will be slightly spread out in all directions uniformly. Thus, the image of a point source formed by a perfect lens having a circular aperture will not be a true point, but it will consist of a small circular spot of light having a finite diameter. As the lens aperture is reduced, the finite diameter of the elementary image will increase until in the limit, with an infinitely small aperture, the light from a point source would spread out indefinitely and cease to form an image at all.

The effective diameter of this "diffraction spot" is found to be about equal to the *F*-number of the lens expressed in micrometers; thus, a perfect *f*/4.5 lens at full aperture would image a star as a tiny disk of light having a diameter of about 4.5 μm. The size is not very definite because the image is bright in the center and fades off around it, with no definite or sharp edge. This star image is extremely small, and it would require a microscope to make it visible to the eye. However, at very small apertures such as *f*/100, the effective diameter of the disk becomes 1/10 mm, which may be quite noticeable.

In ordinary photography it is much more useful to consider the effect of diffraction *at the object itself* rather than at the image formed inside the camera. Suppose we are photographing an object 50 feet away with a 2-inch lens stopped down to *f*/64. The linear lens aperture is now 2/64 = 0.031 inch, and at 50 feet its effective *F*-number at the object is 50 × 12/0.031 = *f*/19400. At this extremely low aperture the diffraction blur has a diameter of about 19 mm, or 3/4 inch. It is therefore impossible with this camera to detect anything smaller than 3/4 inch at 50 feet. A brick wall 50 feet away photographed with this camera would fail to differentiate any bricks, as the mortar is less than 3/4 inch thick, and would appear as a uniform gray surface. To avoid this situation, most modern 2-inch camera lenses are made so that they cannot be stopped down below *f*/16; at that aperture the diffraction disk at 50 feet would have a diameter of about 0.2 inch. With a 28mm lens at *f*/16, the blur circle would be larger, about 0.34 inch.

Of course, large lenses can be used at quite low relative apertures. For instance, an 8-inch lens at *f*/64 has a linear aperture of 1/8 inch; the diffraction blur at 50 feet would have a diameter of about 0.2 inch, which is acceptable.

Our discussion so far has dealt with perfectly corrected lenses. If the lens is imperfect, and all practical lenses have some residuals of aberration, then the size of the star image will be somewhat larger than the value we have assumed. When the lens iris is stopped down from its maximum aperture, the effect of the aberrations will become progressively less, and that of diffraction progressively greater, until at some particular aperture the two effects will be equal. We shall then have the best image sharpness of which the lens is capable. This point often falls at about two stops down from the maximum aperture of the lens.

The Pinhole Camera

The pinhole camera consists only of a tiny pinhole at one end of a darkened box and a film at the other. A narrow beam of light from any point in the object will pass through the hole and fall on the film. Consequently, the larger the pinhole, the larger will be the spot of light corresponding to a single object point (Fig. 3.1). However, if the pinhole is made too small in an effort to achieve better definition, diffraction comes into play, leading once more to a large patch of light on the film. The optimum size of pinhole

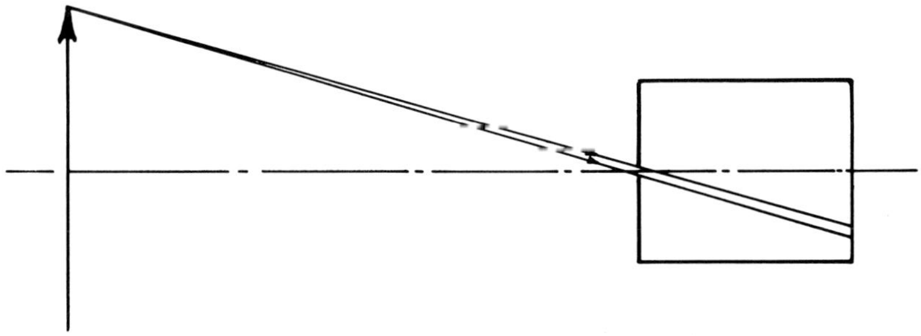

Figure 3.1. The action of a pinhole camera.

thus falls at the crossing point of these two effects, when the diameter of the hole is just equal to the diameter of the diffraction image. This occurs when

$$D = \frac{f}{1000\,D} ,$$

where D is the hole diameter in millimeters. Hence $D^2 = f/1000$, or

$$D = \sqrt{0.001f} = 0.032\sqrt{f} \text{ millimeters} .$$

Some typical values are given in the following table:

Length of camera, f	2	3	4	6	10	inches
	50	75	100	150	250	millimeters
Optimum pinhole, D	0.23	0.28	0.32	0.39	0.51	millimeters
F-number	217	267	312	384	490	

These figures are not at all critical, both by the very nature of the whole argument and because the diffraction image of a point of light is not a uniform disk with a sharp edge but has a bright center that fades out to a rather indefinite extent. Thus, we would not noticeably lose definition if we were to make the hole 20 to 30% larger than the sizes given in the above table, and we should at the same time gain very significantly in light. In some articles* on the subject, the optimum diameter of the pinhole is stated to be $0.041\sqrt{f}$ mm.

The fundamental accuracy of the diffraction approach has been checked many times by direct photography, the classical book by Colson being well worth study by those interested in the subject. A typical pinhole photograph is shown in Fig. 3.2.

The Crossed-Slit Anamorphoser

An interesting device, credited to Ducos du Hauron, is the *crossed-slit anamorphoser*. This is a modified pinhole camera in which the pinhole has been replaced by a pair of narrow, perpendicularly crossed slits spaced apart along the camera axis (Fig. 3.3). The horizontal scale of the picture is

*E. W. H. Selwin, "The Pinhole Camera," Phot. J. **90B,** 47 (1950).
R. Colson, *La Photographie sans Objectif*, Gauthier Villars, Paris (1891).
J. M. Field, "Survey of Pinhole Optimization," J.S.M.P.T.E. **74,** 320 (April 1965).

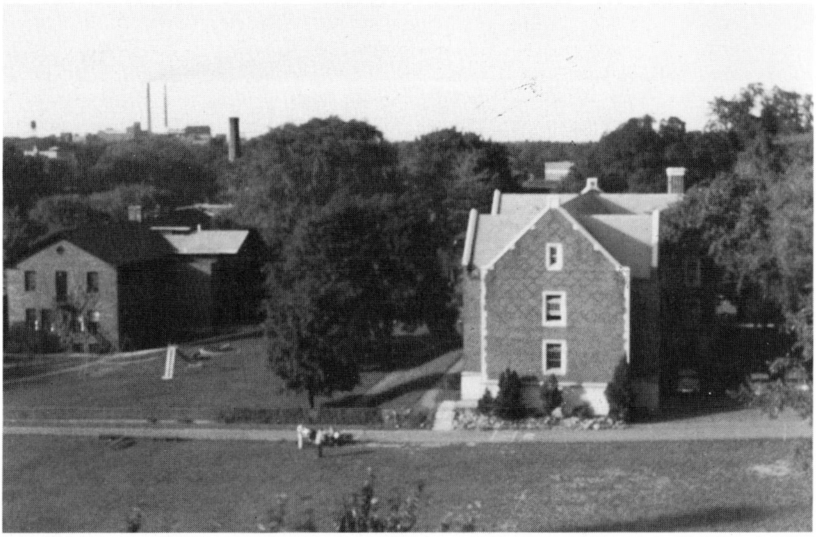

Figure 3.2. A pinhole photograph.

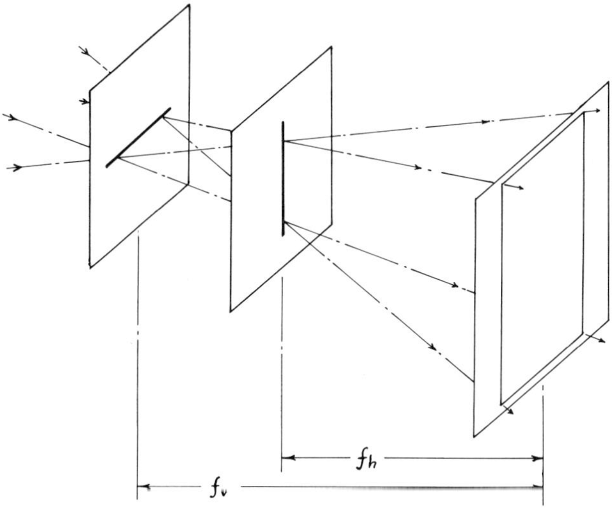

Figure 3.3. The crossed-slit anamorphoser.

obviously determined by f_h, which is the distance from the vertical slit to the film, but the vertical scale is defined by the distance f_v of the film from the horizontal slit. The pair of slits working together thus constitutes a pinhole camera in which the image is stretched or compressed in one direction more than in the other. This type of distortion is called "anamor-

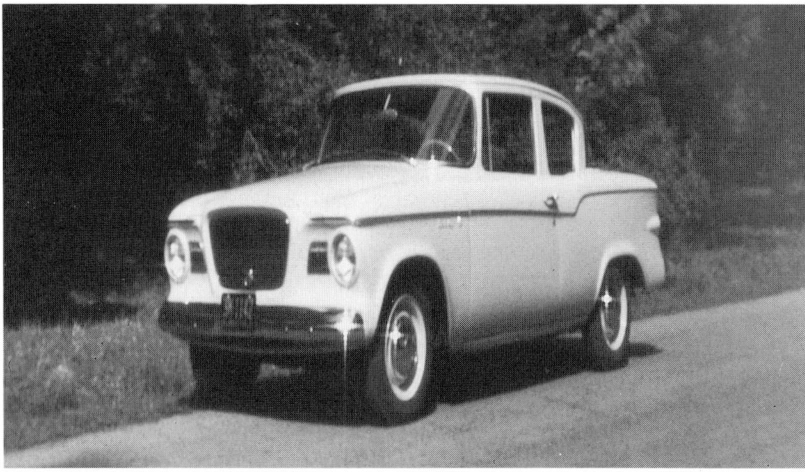

(a)

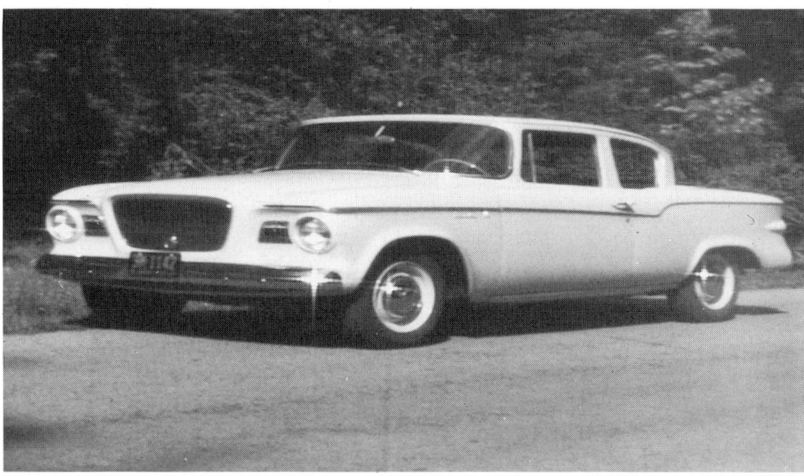

(b)

Figure 3.4. The effect of anamorphic compression.

phic" or "anamorphotic," and the degree of anamorphic compression can obviously be varied over a wide range by merely changing the separation of the slits or by moving the pair of slits closer to or further from the film. Two photographs made with this device are shown in Fig. 3.4. They were taken from the same viewpoint as the lower view in Fig. 1.10, the compression ratio being about 70%. Some other forms of anamorphic attachments are described on page 190.

Chapter 4

Definition and Resolution

Photographic Definition

Many factors affect the definition of a photographic print, and although *definition* in photography is used to indicate the quality aspect of a photograph that is associated with the clarity of detail, no way has yet been devised to evaluate it in other than relative terms. The precise effects of the factors on which it depends, such as resolving power, graininess, sharpness, and contrast, have not been fully determined. Furthermore, the problem of evaluation is complicated by the fact that a subject having fine detail, such as an aerial view of a city, looks unsatisfactory unless the definition is exceptionally good, whereas a close-up of a person may be considered a good likeness even though the detail definition is poor.

Any photographic subject may be thought of as an aggregate of a vast number of single points, and a perfect lens should, of course, image each object point as a point of light correctly located on the film. However, because of the limited powers of the human eye, it is unnecessary that each object point be imaged as a true point, and in practice a small patch of light is indistinguishable from a point if it is smaller than some limiting size. We shall make use of this concept when discussing depth of field (page 84), and we are now applying it to discuss the factors that limit the definition obtainable in a negative or in a positive print.

The Measurement of Resolving Power

Resolution in the photograph has been commonly used as a measure of definition. The resolving power of a photograph is usually expressed as the number of line pairs per millimeter that are just clearly resolved, or separated, by the entire photographic process. The bars in the test object used to measure resolving power are generally bright on a dark background, and they are separated by spaces that are as wide as the bars themselves. It is customary to use a pattern containing three or four bars, but some workers prefer only two. It is desirable to keep the ratio of the lengths to the widths of the bars constant for all the patterns, and the customary arrangement is to have the proportions such that the outline of each group is square, as shown in Fig. 4.1.

The camera is very carefully focused on the test object and then, since resolving power depends markedly on exposure, an exposure series is made. Each negative, or a print from it, is examined carefully with a low-power microscope, and the last group of lines from the coarse end of the series that is still just resolved in the best image is noted [Fig. 4.2(a)]. If a finer group of lines should happen to be resolved, beyond a group that is not resolved, this represents what is known as *spurious resolution* and should be ignored. When spurious resolution is present, the apparent number of black lines in the pattern is often less than the true number; for example, only three lines appear in the image of the sixth set of Fig. 4.2(b).

The selection of a suitable interval between the numbers of lines per millimeter in adjacent charts requires some consideration, for if they are too widely spaced, the limiting line spacing may fall between two groups, and some group thereafter may show spurious resolution. The tendency is to put size groups within a range of 2:1 at $\sqrt[6]{2}$ steps, so that a typical series

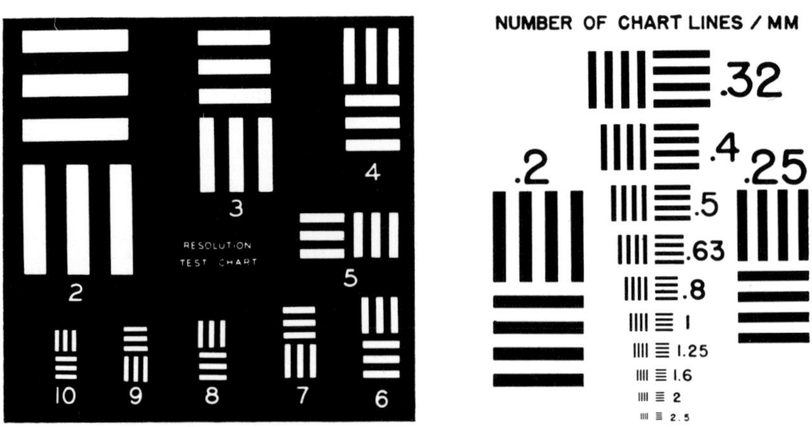

(a) (b)

Figure 4.1. Two typical resolution charts. The numbers in chart (a) represent lines per millimeter when copied at 12× reduction. The numbers in chart (b) represent lines per millimeter on the original chart.

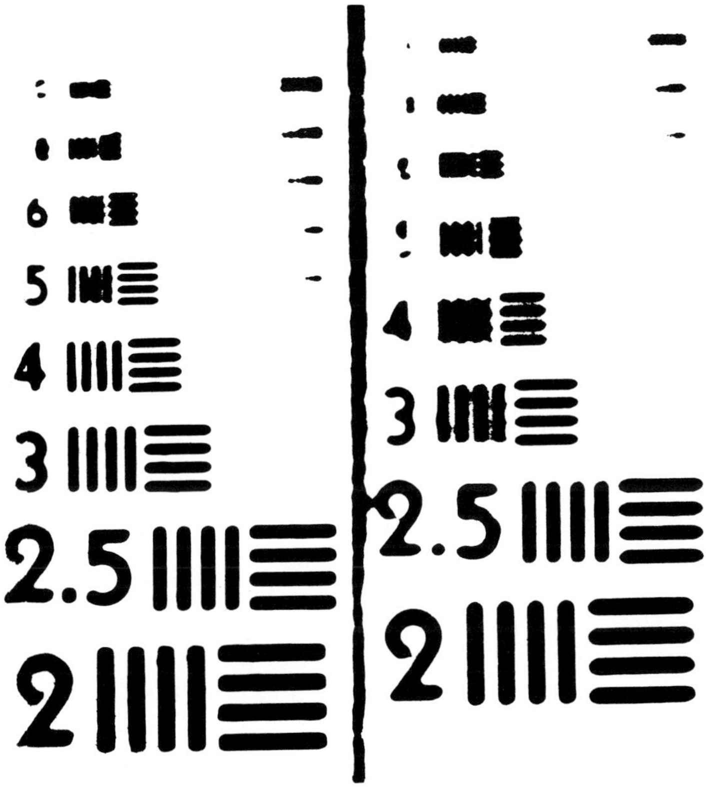

(a) (b)

Figure 4.2. Two typical resolution photographs, much enlarged, taken at different distances from the lens. In (a) chart 5 is still resolved but chart 8 is showing spurious resolution. In (b) chart 3 is the last to be resolved. Chart 5 shows only 3 black lines and chart 8 shows only 2 black lines.

of charts would represent . . . 10.0, 11.2, 12.6, 14.1, 15.9, 17.8, 20.0, 22.4 . . . lines per millimeter in the image.

Experienced lens testers find that, when judging a photograph of resolution charts, the microscope power must be neither too high nor too low. If the magnification is too low, obviously the finer lines will be too small to be seen. If the magnification is too high, the graininess may well become overwhelming and the problem then becomes one of judging whether there are any regularly spaced regions of comparatively dense and sparse grains. However, even if graininess is not a problem, too high a magnification may reduce the density gradient in the image to the point

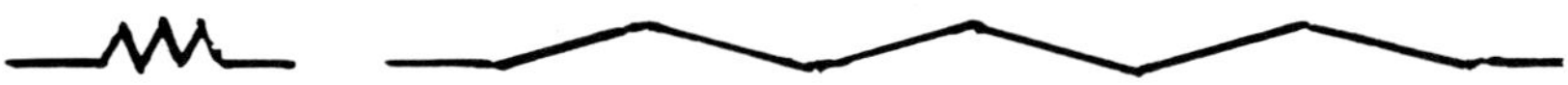

Figure 4.3. The gradient of a bar-chart image: left—at low magnification, and right—at high magnification.

where the lines can no longer be seen. Steep gradients of density are readily visible to the eye, but flat gradients are often invisible, even though the height of the peaks above the valleys is the same in both cases (Fig. 4.3.).

The Criterion of "Good Definition"

Although the influence of the various factors that affect definition has not been studied exhaustively, the resolution in the photograph should be equal to what the eye expects to see under the conditions depicted. For paper prints viewed at the normal distance of a foot or so from the eye, a resolution of about ten lines per millimeter seems to be adequate. This is equivalent to an angular resolution of 1.5 minutes of arc at the eye. If the print is an m-times enlargement, then the negative from which it is made should show a resolution at least equal to 10 lines per millimeter multiplied by the factor m. Actually, even the best enlarger lens will cause some further degradation in the image, and it is unusual to obtain even 10 lines per millimeter in an enlargement.

If the image is projected onto a screen, the resolution in the projected transparency should be such that the resolved lines subtend 1.5 minutes (1 in 2500) at the observer's eyes. For the particular case of 16mm projection with a 2-inch lens, when the observer is situated midway between the projector and the screen, this minimum resolution would correspond to about 100 lines per millimeter on the film. However, because of motion in the picture and many other contributory factors, it is found in practice that a resolution of about 60 lines per millimeter in the positive film gives excellent definition on the screen, provided a good projection lens is used. For 35mm motion picture projectors, a resolution of 30 to 40 lines per millimeter on the positive film would be regarded as excellent definition by a critical observer sitting in the middle of the auditorium.

As a rough guide, we may say that the resolution required in a negative, to yield acceptable definition, is about equal to 200 lines divided by the square root of the long dimension of the negative in millimeters. This formula gives 62 lines for 16mm film, 40 lines for single-frame 35mm film, 33 lines for double-frame 35mm film, and so on down to 15 lines for a 5 × 7-inch negative.

With adequate resolving power, sharpness and graininess become important determinants of definition. When one of these characteristics is constant, the other determines the quality of the definition; when they both vary, it is found that sharpness has a greater influence on definition than does graininess. So far, nothing has been said about the tone characteristics or contrast of the photographs, and it has been assumed that the densities of the corresponding parts of the photographs are the same. The effect on definition of changing the density differences between the bright and the dark parts of the photographs has not been studied.

Modulation Transfer Function*

From what has been said, it is clear that neither resolution, sharpness, graininess, nor tone reproduction is of itself the determining factor for good definition in a photograph. In order to combine these properties in the simplest way, it has recently become customary to express the performance of a lens by its *modulation transfer function* (MTF), on axis and at several points in the field.

As its name implies, MTF is a measure of the ability of the lens to form an image that is an accurate reproduction of an object. Suppose we use the lens to form an image of a sequence of bright and dark bars having a sine-wave distribution of luminance across the bars, with a gradually decreasing separation between successive bars, as shown in the top row of Fig. 4.4. (The reason to require a sine-wave cross-section of the bars is because the image of a sine wave is always a sine wave, no matter how good or how bad the lens may be.) The image of this series of bars, shown in the middle of Fig. 4.4, is then scanned by a narrow slit parallel to the bars, behind which is mounted a sensitive photocell to record the amount of light passing through the slit. The output from the photocell as the slit moves along the image is shown by the trace at the bottom of Fig. 4.4. When the bars are broad and widely spaced, the lens forms an image that is a good reproduction of the object; but as the lines become narrower and closer together, light from the bright bars spills over into the dark bars, thus reducing the brightness of the bright bars and raising the brightness of the dark bars, until eventually all the bars become equally bright and resolution is lost.

From this trace it is possible to determine the *contrast* of the image, defined as the intensity ratio (bright minus dark)/(bright plus dark) at each

* F. H. Perrin, "Methods for appraising optical systems," J.S.M.P.T.E. **69**, pp. 151, 239, 800 (1961); G. C. Higgins, "Methods for engineering photographic systems," Appl. Opt. 3, 1 (1964).

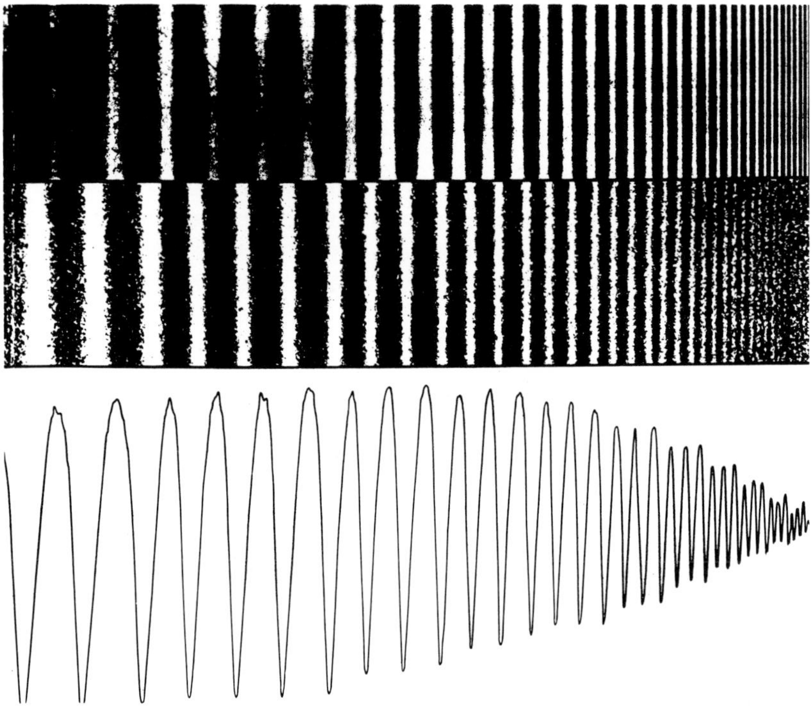

Figure 4.4. The meaning of MTF. (Top) A sine-wave bar chart with progressively closer lines. (Middle) A photograph of the sine-wave bar chart. (Bottom) A microdensitometer trace of the photographic image.

line frequency. It is obvious from this relationship that if the dark lines in the image are perfectly black, the contrast will be 1.0, and if there is no difference between bright and dark bars, the contrast will be zero. A plot of contrast versus line frequency, expressed as so many lines per millimeter in the image, is called the *modulation transfer function* of the lens for the particular conditions of measurement. The MTF curve differs from one point to another in the field, and it also differs depending on the orientation of the lines, whether radial or tangential, with the wavelength of the light, and the position of the image plane along the axis. A complete test of a lens, therefore, results in a stack of MTF graphs that is sometimes difficult to interpret but which contains extensive information about the lens performance.

Figure 4.5 shows a typical set of MTF curves for a photographic lens, taken at four different field angles. It will be seen that in this particular lens, there is virtually no resolution beyond about 40 lines per millimeter.

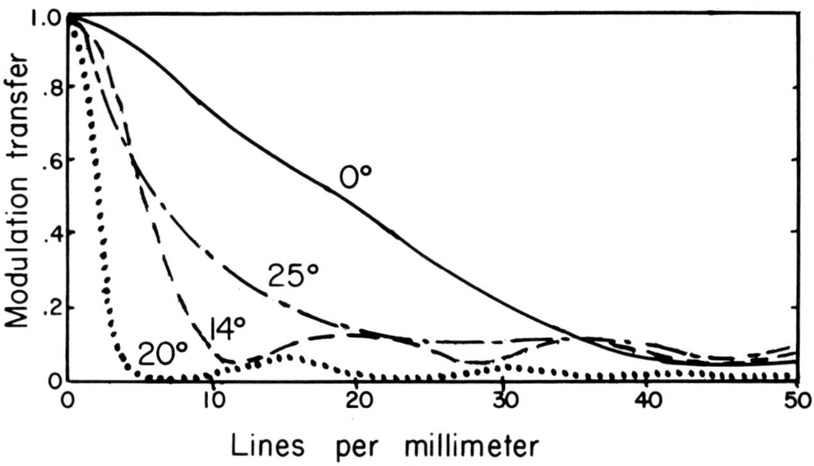

Figure 4.5. A plot of MTF at different points in the field of a typical lens.

These curves were plotted for full aperture. On stopping the lens down, however, greatly improved imagery will be obtained.

The MTF curves of a lens can be plotted from theory if the structure of the lens is known, or they can be measured in the laboratory if the lens itself is available. Lens designers are becoming increasingly dependent on calculated MTF values to determine if a design is likely to be good enough for some particular application.

The use of MTF as a quality criterion has an important additional virtue, namely, the possibility of *cascading* the MTF of a lens with the MTF of a film or other diffusing surface in the image plane. The MTF of a combination of lens and film can be found by multiplying the separate MTF values of lens and film at each spatial frequency. Note that the MTFs of two lenses used in succession cannot be cascaded unless there is a diffusing surface between them. However, we *can* cascade the successive MTFs of a camera lens, a negative film, a printer lens, a positive film, a projection lens, and a screen, because each lens is then separated from the next by a diffusing surface. When this is done, the result is the relation between the contrast of the original scene and that of the final projected image on the screen.

Causes of Poor Image Definition

There are many stages in the photographic process at which some loss of definition may occur, due both to the equipment and also to how it is used, and the photographer who is interested in getting the highest possible definition will do well to bear these in mind. The principal sources of poor definition may be summarized as follows, in the order in which they are most likely to cause trouble:

(a) the picture-taking process
(b) the camera
(c) printing (or projection)
(d) the film
(e) development of the film.

Each of these causes will now be considered.

The Picture-Taking Process

Even if the camera itself is in perfect adjustment, the actual operation of taking a picture is full of sources of poor definition. Movement of the camera during exposure by careless operation of the shutter release, particularly when using a long-focus lens, is an obvious hazard. Conversely, use of a short-focus lens makes camera movement, or jiggle, less serious. Pictures made with a camera mounted on a rigid tripod are generally much sharper than those made with a hand-held camera. A heavy camera is less susceptible to jiggle than a lightweight one. Short exposures show much less jiggle than long exposures, and for this reason simple automatic cameras now have exposures of 1/80 s instead of the 1/40 s that was common several years ago. Some people hold their breath at the moment of shooting to minimize camera movement.

An easy test of your steadiness in holding a camera is to attach a small mirror to the front of the camera, and place a slide projector so that its light beam strikes the mirror and forms a spot of light on a wall 10 to 20 feet away. Careless shutter operation will cause the spot of light to move.

Inaccurate focusing of the camera, or poor judgment of the object distance, can degrade picture definition. Errors in judging distance can be minimized by stopping the lens down, or by using a small camera with a short-focus lens, since both these means tend to increase the depth of field. Definition can be lost indirectly by having the principal subject occupy so small an area of the negative that excessively great enlargement is necessary in printing. Another indirect reason for poor ultimate definition is a serious degree of overexposure or underexposure, since the resolving power of the emulsion is low under such conditions. The mechanical inertia of the moving parts in the shutter is a possible cause of lost definition, but this is not usually very serious.

Some of these troubles are aggravated when a very small or miniature camera is used. Its light weight may tend to increase the risk of camera movement at the instant of trigger release, and the increased enlargement required in printing makes defects of all kinds very prominent. However, focusing becomes far less sensitive in such a camera, as the depth of field at a given object distance is inversely proportional to the linear diameter of the lens aperture.

Finally, movement of the subject itself during exposure should not be overlooked as a possible cause of poor definition.

The Camera

The lens and its aberrations are an obvious fundamental source of poor definition, and even if the lens is so well designed that its aberrations are theoretically negligible, it is entirely possible, owing to errors in manufacture, that some lack of centering or other defect may be present that will cause poor definition. Lens testing is a difficult art requiring special apparatus, but a visual test with either a distant point source or an extended source such as a page of print, using a compound microscope to study the aerial image, will tell a great deal about the lens quality.

A warning must be given here against looking at an aerial image with a low-power magnifier without any ground glass in the focal plane. This is likely to make the image look much better than it should, because the eye is probably using only a small portion of the beam of light from the lens being tested. This is equivalent to stopping the lens down to a small aperture (*EE′* in Fig. 4.6). On the other hand, a compound microscope of 50 or 100 power has such a small exit pupil that there is little likelihood that the observer's eye will isolate a small part of the emerging beam. Visual resolution figures determined in this way with a microscope are much higher than the best that can be obtained in the camera, because the film also contributes its share to the net resolving power of the lens-film combination.

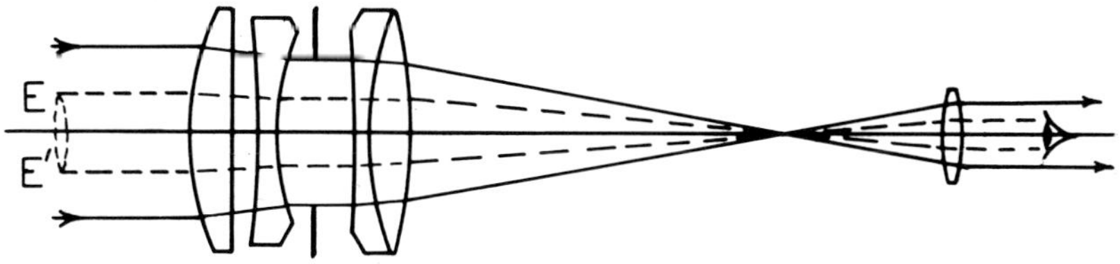

Figure 4.6. The use of a low-power magnifier to examine an aerial image.

Resolution usually deteriorates fairly quickly as we move further out into the lens field away from the axis, owing to the oblique aberrations of the lens, such as field curvature, coma, and astigmatism. When astigmatic aberrations are present, the resolving power for radial lines is different from the resolving power for tangential lines. Care should therefore be taken to ensure that the test chart contains both radial and tangential lines, and it is usual to state whichever resolution is the *less* at each obliquity. The variation in resolving power with obliquity and position of the focal plane along the axis for a typical lens is shown in Fig. 4.7. The locus of the peaks of the resolution curves gives a good idea of the field curvature of the lens.

In this connection it is worth remarking that the edge separating a bright area from a dim area may appear unsharp on a print yet quite sharp in a projected color slide. This is due to the increased brightness of a projected image, so that if the width of the blurred edge is the same in the

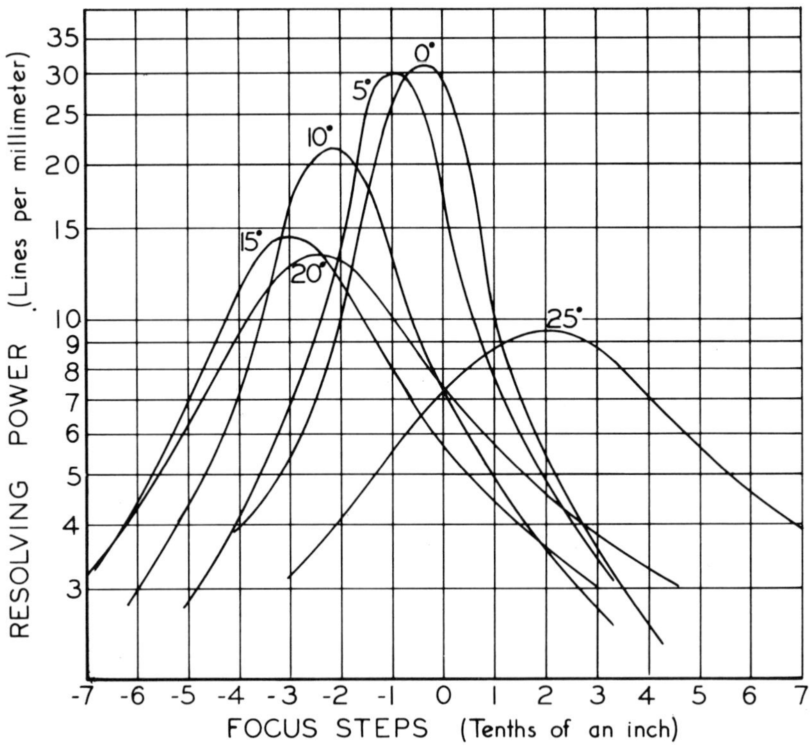

Figure 4.7. The variation of resolving power with position of the film plane at several different obliquities. The lens used was a typical 19-inch process lens at *f*/11 and unit magnification.

two cases, the *gradient* of illumination across the edge will be much steeper in projection, and thus the edge appears sharper to the eye.

A related phenomenon is concerned with the apparent blackness of a thin line seen on a white background. If the width of the line is comparable to the width of the visual blurring, the contrast will be reduced and the line will appear gray rather than black. However, if the line is viewed through a lens, it appears much blacker because the magnified line image is now wider than the width of the visual blur.

The Diffraction Limit of Resolution

The term resolution has long been used to refer to the clear separation of close parallel lines in the image formed by an optical system. If the object consists of a set of parallel, narrow, bright lines on a dark background, the spaces being equal in width to the lines, it has been found both theoretically and experimentally that the lines will just be resolved by a perfect lens if the number of lines per mm is equal to about 1600 divided by the *F*-number of the lens. This figure is the so-called "diffraction limit" of resolution, and if a lens actually reaches this theoretical limit of resolution it is said to be "diffraction limited." (We are referring here to the aerial image formed by the lens and not to the image recorded on film.) Of course, in any practical lens the inevitable residuals of aberration are generally great enough to depress the resolving power to a point drastically lower than the diffraction limit, and diffraction-limited lenses are extremely rare.

Acutance and Sharpness

It is common experience that the boundaries between light and dark areas appear sharper in some photographs than in others. These differences depend mainly on the structure of the image of an object-point formed by the camera and enlarger lenses, but they also depend somewhat on differences in the light-scattering properties of the various emulsions. Although both sharpness and resolving power are determined ultimately by the properties of the lenses and the emulsion, they are quite different phenomena, and a sharp photograph may have low resolving power, and vice versa. A new term, "acutance," has been applied to a quantity that can be determined in the laboratory and serves as a measure of sharpness.

The mounting of the lens and the film in the camera may cause trouble and should be checked if there is an appreciable loss of definition. The lens axis may not be precisely perpendicular to the film plane, or one of the lens elements may be slightly tilted; each of these errors will result in a tilted field. At the image plane the film may be bulging or buckling for some reason, causing some areas of the negative to be less sharp than others. If the camera has a ground-glass focusing screen, care should be taken to check that it is properly in register with the film plane.

It is worth noting that when rolling a paper-interleaved film through the camera, as is the case when 120, 620, or 126 film is used, the camera must be carefully designed so that the strip of paper never becomes tight while the film is slack, as this would lead to a serious buckling of the film. It is better if the film becomes tight while the paper is slack. Of course, this is no problem if the film is in a cassette with no paper backing.

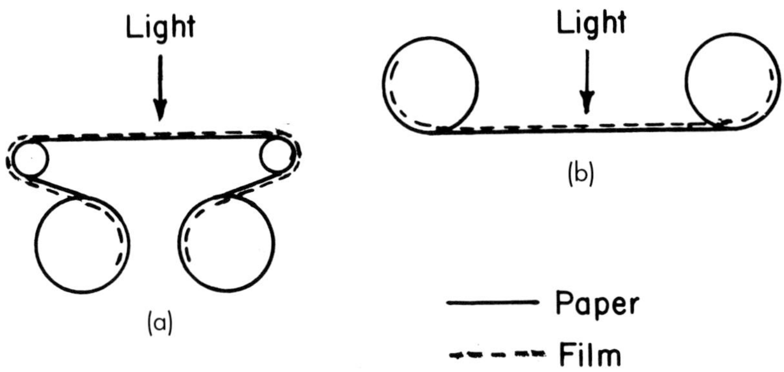

Figure 4.8. Two ways of mounting a paper-backed roll film in a camera.

Another possible cause of film buckle is when the film, which is sold tightly rolled with the emulsion side in, is bent back on itself in the camera, as in Fig. 4.8(a). Here the camera can be quite compact, but the film has to reverse its curvature; it naturally tends to bulge upward toward the incoming light and it must be held flat in the camera. In the other arrangement, Fig. 4.8(b), the film follows its natural curvature, but the camera has to be made larger to hold the supply and take-up reels.

The accuracy of the focusing scale, and of the range finder if one is provided, have an obvious influence on the quality of the definition, and these should be checked if the best definition is required. Since a coupled range finder involves cams and linkages that are delicate and may easily get out of adjustment, it is advisable to test the focus scale and range finder separately against a piece of fine ground glass laid against the film gate in the camera, using a magnifier to view the image on the ground glass. The scale and range finder settings may disagree slightly, but no harm will result if the error is less than the depth of field of the camera under the test conditions.

Even if camera and lens are perfect, the quality of the picture will be seriously affected if there is, for example, a thumbprint on the face of the lens. Small bubbles in a lens are quite unimportant since they do not in any way impair definition. Indeed, at one time the presence of bubbles was considered an advantage as it could be regarded as an indication that special types of optical glass had been employed by the lens manufacturer to give the highest possible definition!

The effects of *diffraction* on resolution should not be overlooked. As explained in the preceding chapter, diffraction may cause a significant loss of definition at very low lens apertures, and the ultimate limit of resolution of any lens when all the aberrations have been eliminated is also set by diffraction.

Printing

A contact print is likely to reproduce nearly all the definition that is in the negative provided the contact between negative and print material is good. Even a small airspace between the paper and negative can lead to a drastic reduction in print sharpness.

Enlargements may suffer in definition because of the aberrations and diffraction properties of the enlarging lens, which are fundamentally similar to those of a camera lens. If an enlarger lens is stopped down very far, diffraction will cause a considerable softening of the print definition, but if it is used at its maximum aperture the lens aberrations may be significant and very careful focusing becomes necessary. The photographer must therefore judge how far to stop the enlarger lens down. A safe rule is that the linear aperture of the lens should not be less than 1/100 of the lens-to-paper distance.

Great care is necessary in focusing an enlarger. The user is at a fundamental disadvantage here in that one focuses the enlarger by dim illumination and views the resulting print in bright daylight. A *slight* error in enlarger focus may lead to a considerable reduction in the graininess of the print, but loss of definition will be an invariable accompaniment. Enlarger vibration should not be overlooked as a possible cause of poor print definition.

Projection

Small color films and motion pictures are, of course, projected onto a screen for viewing. In this case, the greatest care must be taken to focus the image accurately and to check the focus after the projector has been in operation for a while. Color slides in cardboard mounts often buckle slightly under the heat of the projection lamp and may require refocusing after a few seconds of showing. Motion picture projectors for 8- or 16mm film frequently require refocusing during a show because the metal parts of the projector slowly heat up and expand, thus moving the lens slowly away from the film. The depth of focus of an *f*/1.6 lens used for projecting 16mm film is only about a thousandth of an inch, and the thermal expansion of the projector during the first half-hour of running can easily amount to several times this value.

The projectionist is generally in a worse position to judge image sharpness than anyone else in the auditorium because he or she is the furthest from the screen. Advice from someone near the front is not conveniently given. Consequently, the projectionist is strongly urged to employ a low-power telescope or opera glass for aid in focusing. The whole pleasure in watching a motion picture can be lost if the screen picture is even slightly out of focus.

The Film

The resolving power of an emulsion is measured in the laboratory by forming a carefully focused image of a resolution chart upon it, by means of a very well-corrected lens such as a microscope objective. A series of exposures of varying length are made, and a graph is plotted connecting the limiting resolution on the film, as seen through a low-power microscope, with the density in the image of the white parts of the original chart. This graph rises to a flat peak at some intermediate density, usually between 1.0 and 1.5, and drops off when the density is either much lower or much higher than the optimum. If other charts are used in which the object contrast is much lower, a series of resolving-power curves can be similarly plotted, as shown in Fig. 4.9. Furthermore, the entire family of curves will change if the development conditions are changed, so it is evident that no single number can be chosen that will adequately represent the resolution properties of any emulsion, any more than a single number can be used to express the resolving power of a lens. The figure given by the manufacturer is the highest resolving power that it is able to attain with clear lines on an opaque background, when the density is optimum and with recommended processing conditions for that particular emulsion. Such high resolution would seldom be attained in normal use of the film in everyday photography.

The resolving power of emulsions is limited chiefly by two factors, namely, the *diffusion* of light, arising from the turbidity of the emulsion, and

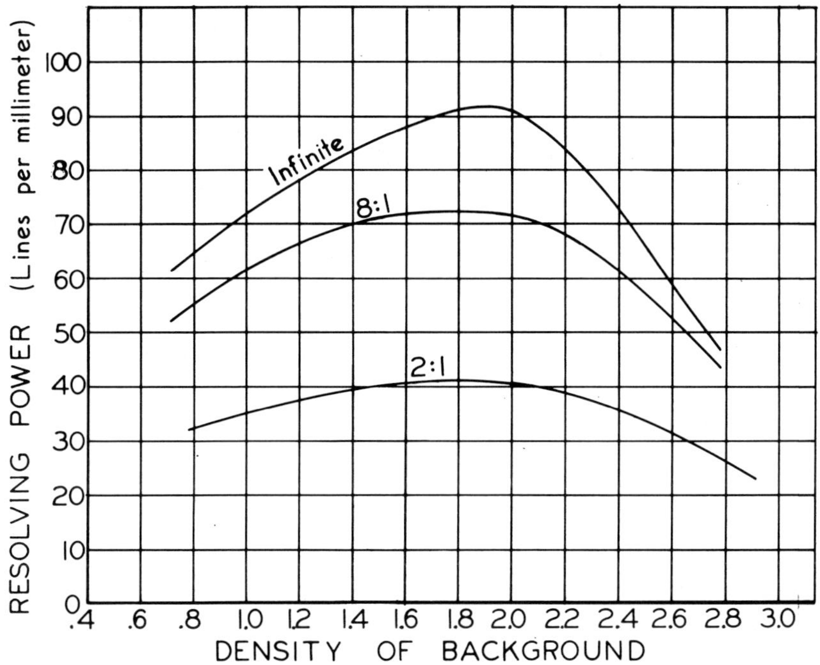

Figure 4.9. A typical family of resolving-power curves for a rapid negative emulsion and three different values of the test-object contrast.

the *graininess* of the developed image. The diffusing properties of the emulsion cause light from a bright line in the image to be scattered into the neighboring part of the emulsion under a dark line, which should not be receiving any exposure. Diffusion thus tends to prevent resolution when the lines in the image are too close together, particularly at low object contrast. Graininess, on the other hand, tends to prevent the observer from being able to say whether or not the images of adjacent bright lines are really separate or whether they have merged together. Thus, in the absence of noticeable graininess, the observer would consider the emulsion to be capable of resolving sets of lines somewhat closer than those actually resolved. The whole subject of resolving power of emulsions is quite complicated, and the reader is referred to other sources* for a more extensive discussion.

* See, for example, C. E. K. Mees, *The Theory of the Photographic Process*, pp. 1011–1024, Macmillan, New York (1954); C. N. Nelson, "Safety factors in camera exposures," J.S.M.P.T.E. **69,** 479–483 (1960).

It should be added that graininess increases with density, while acutance and resolving power diminish beyond a certain maximum, and therefore the camera exposure should be as short as will result in good tone reproduction. A low density is especially desirable for negatives that are to be enlarged, not only to keep the exposure time in the enlarger short, but also because a higher magnification is possible without obtrusive graininess.

Combination of Lens and Film Resolution

The problem of expressing the relative contributions of lens and film to the net resolving power of a complete camera depends on MTF measurements, but it appears that in some cases the resulting resolution may be found by reciprocal addition; thus,

$$\frac{1}{\text{net resolving power}} = \frac{1}{\text{lens resolution}} + \frac{1}{\text{film resolution}} ,$$

all resolutions being expressed in lines per millimeter. Thus, if the resolution of the lens is just disappearing at 100 lines per millimeter (visual) and the film is capable of just resolving 50 sharp, clear-cut lines per millimeter, then the combination of lens and film may be expected to have a resolving power of perhaps only 30 lines per millimeter. However, this simple rule makes no claim to great accuracy, and there may be many cases to which it would not apply.

Development of the Film

The type and time of development have little effect on resolving power. Fine-grain developers often improve resolving power but sometimes reduce acutance. Prolonged development increases graininess but also increases contrast; however, since a contrasty negative calls for a soft paper, the increase in graininess is partially neutralized. When a negative is to be used for a 10× enlargement, graininess must be considered, and therefore a fine-grain film should be used. The graininess of a fast film can be reduced by using a fine-grain developer, but usually at the cost of a loss of speed, and, with some developers, of sharpness, also.

Large Versus Small Cameras

Briefly, large cameras gain in definition but lose in depth of field. Small cameras gain in portability, lens aperture, and depth of field, but lose in definition unless great precautions are taken. Small cameras also gain in cost of film, which becomes important in color photography. Indeed, many professionals resort to 35mm cameras only for color shots, using a larger camera for black and white. The larger sizes facilitate retouching and the identification of the subject by direct viewing of the negative. Large negatives yield very sharp prints, even at camera apertures as low as *f*/64.

Before the days of enlargers, which means before the invention of convenient artificial-light sources, the size of the print was necessarily the same as the size of the negative, and 8 × 10-inch or 6½ × 8½-inch were normal routine sizes. By the time of the first World War, 4 × 5-inch and 5 × 7-inch had become common; by 1930, 3¼ × 4¼-inch and 2¼ × 3¼-inch were standard, and the 24 × 36mm of the Leica was regarded as a "miniature" size. By 1940 few amateurs used negatives larger than 2¼ × 3¼ inches or 2¼ inches square, but by 1950 the 2¼ × 3¼-inch size had gone, and 1⅝-inch square on 127 film was becoming popular. By 1960, 2¼-inch square was regarded as the largest amateur size, 1⅝-inch square and 24 × 36mm were standard, and a series of smaller sizes on 35mm or 16mm film were appearing.

The square 126 film size (26 × 26mm) appeared in March 1963, the smaller 110 size (13 × 17mm) early in 1972, and the Disc camera in 1982. In spite of its small picture area, 110 film has become enormously popular, and cameras using that film have been made by most camera manufacturers. These small formats would not have been acceptable with the older grainy and low-resolution emulsions, but today they are usable even at high ASA speeds. One can argue that 16mm and 8mm motion pictures use even smaller frame sizes, but the secret there is that at 16 or 18 frames per second the grain pattern changes so rapidly that the grain is never visible to the eye. A single frame of these films is not satisfactory.

Chapter 5

Depth of Field

It is, of course, well known that if we focus a camera on an object at some definite distance from the camera lens, there will be a finite range of distances in front of and beyond the focused object in which everything appears acceptably in focus, while outside that range everything becomes progressively more blurred at increasing distances from the plane of best focus. The actual extent of this range of acceptably sharp definition depends mainly on the distance of the subject from the lens, the aperture of the lens, and on the manner in which we look at the final print, but it also depends to some extent on the type of subject being photographed, the resolving power of the film and paper emulsions, and the aberrations of the camera lens. However, for the sake of simplicity, the following discussion will be based on the assumption that we are using grainless film and aberration-free lenses, and we shall end with a few remarks on the effects of these neglected factors in actual practice.*

For ordinary fairly distant objects, the depth of field of most cameras is generally adequate, provided that we take care to focus the lens on the subject of principal interest. In a portrait this is generally the front of the face; failure to focus correctly in such a case may be artistically disastrous. However, when we have occasion to focus a lens down to one or two feet, we are surprised to find how little depth there is, even at small apertures. For example, a 2-inch lens at *f*/8 when focused on objects at 12 inches distance has a depth of field of about half an inch in front of and beyond the focused distance. A small depth of field is sometimes very useful, for instance, as a means of eliminating an unwanted background (Fig. 5.1).

---

*Much useful information regarding depth of field is given in the following sources: A. Cox, *Photographic Optics*, 15th edition, pp. 68–97, Amphoto, Garden City, N.Y. (1974). S. F. Ray, *Applied Photographic Optics*, pp. 180–193, Focal Press, London (1988).

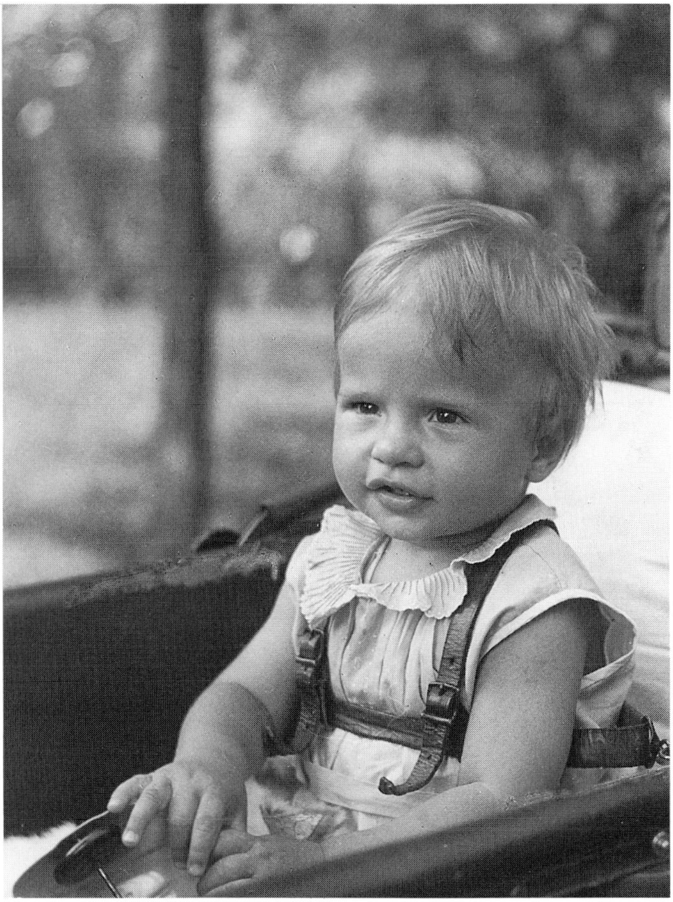

Figure. 5.1. A small depth of field may be used to eliminate an unwanted background.

In order to derive formulae with which the depth of field of a lens can be calculated in any particular case, we must first standardize the manner in which the final picture is to be viewed, since obviously a slight blurring of out-of-focus objects may be quite invisible to someone standing at the other side of the room and yet become distressingly evident when the print is examined closely. As this factor is by far the most important in any depth calculation, we shall treat it in three different ways: (a) we can take into account the actual distance of the observer from the final print or projected image, (b) we can assume that the final picture will be observed from the correct center of perspective, or (c) we can adopt a fixed circle of confusion on the film, the value of which will, of course, depend on the dimensions of the negative size.

We must always remember that there is no abrupt limit to the depth of field. The blurring of an out-of-focus image occurs gradually, and, depending on the position of the viewer in relation to the image, the viewer will begin to realize that some portions of the image are not sharply imaged while other portions are quite sharp. Hence, any definite data on depth of field must be understood as the beginning of perceptible blurring and not an abrupt division between a sharp image and a blurred one.

(a) When the Observing Distance is Known

As an example, we will suppose that the observer of a projected color slide is located at twice the picture width from the screen. We will also suppose that the observer's eyes are incapable of distinguishing between a true point and a tiny blur circle if the circle diameter subtends an angle less than 3.4 minutes of arc (1 in 1000) at the observer's eye. Then the diameter of this limiting circle on the screen will be $2W/1000$, where W is the screen width. As the picture area on a 35mm film slide is 1×1.5 inches, the corresponding acceptable circle of confusion on the film will have a diameter of $2 \times 1.5/1000 = 0.003$ inch or 0.076 mm. Assuming that the original subject was 10 feet from the camera when the picture was taken, using a 2-inch lens at *f*/2, the projected diameter of the circle of confusion on the original object plane would be equal to $0.003 \times 120/2 = 0.18$ inch.

We now refer to the depth-of-field diagram given in Fig. 5.2. Here d is the diameter of the lens aperture (focal length divided by the *F*-number)

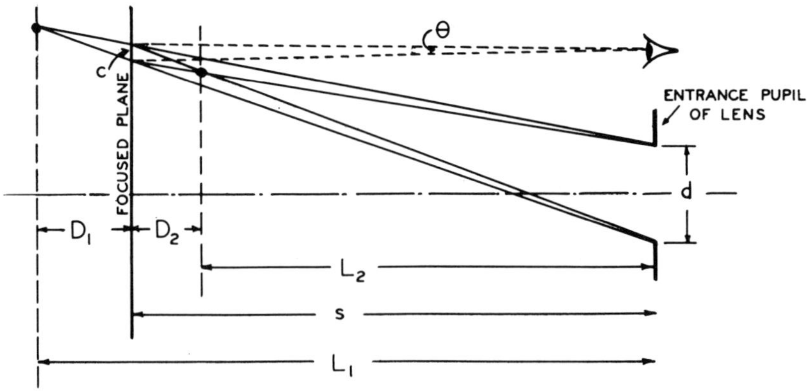

Figure 5.2. Situation in the object space.

and c is the diameter of the acceptable circle of confusion *in the object plane*. If D_1 and D_2 are the far and near depths of field, respectively, it is clear by simple proportion that

$$\frac{c}{d} = \frac{D_1}{s + D_1} = \frac{D_2}{s - D_2} , \tag{5.1}$$

where s is the distance of the focused plane from the lens. Using these relations, we find that

$$D_1 = \frac{sc}{d - c} \quad \text{and} \quad D_2 = \frac{sc}{d + c} . \tag{5.2}$$

The near and far depth limits are at distances from the camera given by

$$L_1 = s + D_1 = \frac{sd}{d - c} \quad \text{and} \quad L_2 = s - D_2 = \frac{sd}{d + c} . \tag{5.3}$$

Inserting into these equations $s = 120$ inches, $c = 0.18$ inch, and $d = 1$ inch, we find that

$$D_1 = 26.3 \text{ inches} \quad \text{and} \quad D_2 = 18.3 \text{ inches.}$$

Hence, any object lying within these D_1 and D_2 distances will not appear detectably unsharp to our assumed observer looking at the projected picture.

This is the only valid way to calculate depth of field, but of course it is not very practical as we never know how the final picture will be viewed.

Size of the Acceptable Angle of Confusion

One may well question the choice of 3.4 minutes of arc as the tolerable size of the acceptable angle of confusion for an average observer. Many careful experiments have shown that the average eye can resolve two close black lines on a well-illuminated clear background, the width of the lines being equal to the space between them, if they subtend at the eye an angle of about 1.5 minutes of arc (1 in 2300). In everyday life, however, these ideal conditions are rarely encountered. The detail in an ordinary photograph is generally of low contrast, and the picture will usually be too dimly lit to allow the eye to exhibit its maximum resolving ability. Hence, it has become customary in depth-of-field theory with small cameras to assume that the eye is unable to distinguish between a true point on a photographic print and a small circle of confusion subtending an angle of about 1 in 1000. This is 3.4 minutes of arc and represents easy resolution for equally spaced black and white lines; indeed, many designers believe that a more stringent criterion should be adopted for large lenses in which the resolving power of the film itself does not affect the depth of field to a significant extent.

As a point of reference, we might note that a dime (18 mm diameter) subtends an angle of 1 in 1000 when viewed at a distance of 59 feet, and so does a pinhead (1.6 mm diameter) at a distance of 63 inches.

(b) When the Picture is Viewed from Its Center of Perspective

Logically, every picture should be observed from its center of perspective, but as this position is not known to the viewer it becomes rather an academic situation. To be sure, we can sometimes locate the approximate center of perspective of a picture by moving toward or away from it and trying to decide when it looks most natural to our eyes, but this is at best a dubious procedure. However, if we can make the assumption that the picture will be viewed from its center of perspective, the calculation of depth becomes greatly simplified.

The center of perspective of the original scene is, of course, at the camera, and if we assume that the observer cannot distinguish between a true point and a circle of confusion subtending 1 in 1500 at the viewer's eyes, then the circle of confusion in the focused object plane becomes simply $c = s/1500$. Thus, in our example of a 2-inch $f/2$ lens with an object 10 feet away, $s = 120$ inches and $c = 0.08$ inch. The two depth values then become

$$D_1 = 10.4 \text{ inches} \quad \text{and} \quad D_2 = 8.9 \text{ inches}.$$

We can simplify the equations still further by inserting $c = s/1500$ in Eqs. (5.2), giving

$$D_1 = \frac{s^2}{1500d - s} \quad \text{and} \quad D_2 = \frac{s^2}{1500d + s} \quad . \tag{5.4}$$

Hence,

$$L_1 = s + D_1 = \frac{s}{1 - \left(\dfrac{s}{1500d}\right)} \quad \text{and} \quad L_2 = s - D_2 = \frac{s}{1 + \left(\dfrac{s}{1500d}\right)} \quad . \tag{5.5}$$

Since the value of D_1 is necessarily larger than D_2, the far depth of field should always be greater than the near depth. However, experiment often fails to bear this out. One reason appears to be that since distant objects in general appear small on the negative while near objects appear relatively large, we can detect blurring due to out-of-focus for distant objects much more readily than for near objects, and this tends to make the far depth appear shorter than the near depth in some cases. We shall ignore this phenomenon in working through the mathematical derivations in this chapter.

In particular, we note that if the picture is to be viewed from its center of perspective, then the depth of field depends *only* on the distance of the subject, the linear diameter of the lens aperture, and the assumed angular resolving power of the viewer's eyes. It does not depend on the focal length or relative aperture of the lens, but only on the diameter of the entrance pupil, equal to the focal length divided by the F-number.

The validity of this statement can be seen in the examples shown in Figs. 5.3 and 5.4. In Fig. 5.3 it is seen how the depth of field varies as the object distance is changed, and in Fig. 5.4 it is clear that the depth depends only on the linear aperture of the camera lens and not on its focal length, provided that the smaller image is enlarged to the same size as the larger image and both are viewed from the same distance. This is one of the principal reasons for the popularity of small film formats, namely, that we can have a large depth of field combined with a fast lens. This is impossible when a large format is used.

The Hyperfocal Distance

It should be noticed from Eq. (5.4) that the far depth D_1 will become infinite if the camera is focused on a distance s equal to 1500 times the lens aperture, assuming that the observer's eyes can resolve an angle of 1 in 1500. This critical object distance h is known as the *hyperfocal distance*. When a camera is focused on this distance, $D_1 = \infty$ and $D_2 = h/2$. Thus, the range of acceptably sharp objects will run from half the hyperfocal distance to infinity. This is the most desirable distance on which to focus the lens on a fixed-focus camera.

When the image is observed from its center of perspective, our fundamental formulae for depth of field can be conveniently expressed in terms of the hyperfocal distance, as follows. From Eq. (5.4) we see that

$$D_1 = \frac{s^2}{h - s} \quad \text{and} \quad D_2 = \frac{s^2}{h + s} , \tag{5.6}$$

and hence

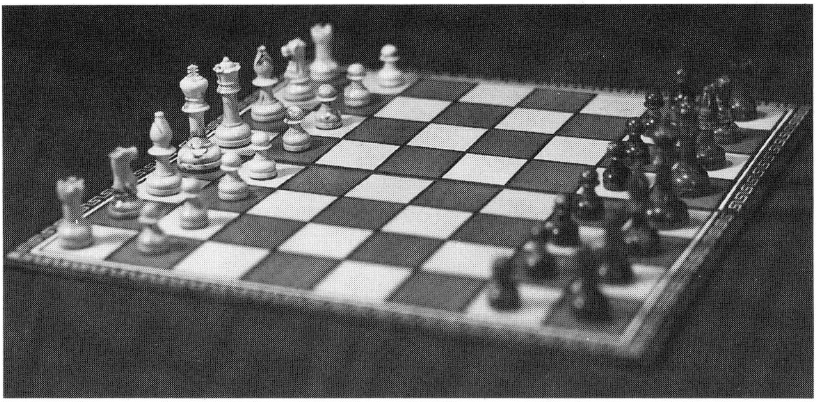

(a)

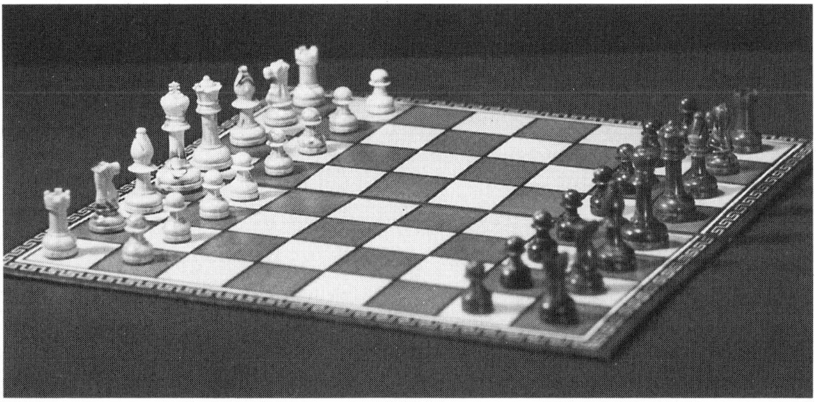

(b)

Figure 5.3. Depth of field varies as the square of the focused object distance: (a) camera focused at 4 feet, and (b) same camera focused at 8 feet, the image enlarged 2× in printing, showing four times the depth.

$$L_1 = s + D_1 = \frac{sh}{h - s} \quad \text{and} \quad L_2 = s - D_2 = \frac{sh}{h + s} \; . \tag{5.7}$$

For relatively close objects we may neglect s in comparison to h, and then each of the depth expressions reduces to

$$D_1 = D_2 = \pm \frac{s^2}{h} \text{ approximately } . \tag{5.8}$$

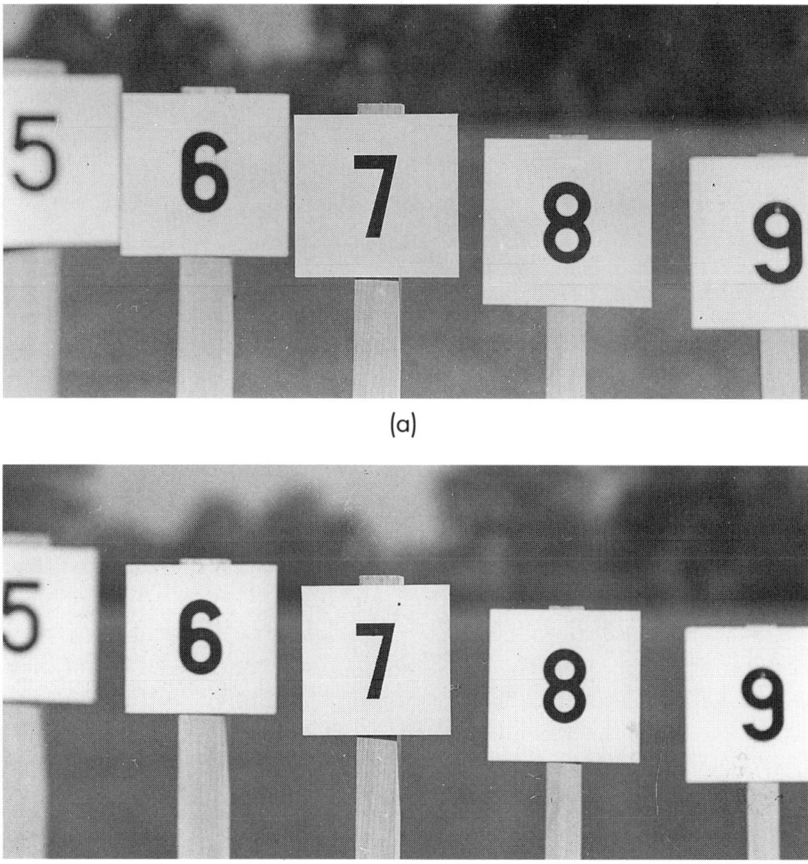

Figure 5.4. Distance targets photographed from the same point with two different lenses having the same linear aperture, namely, 3/4 inch. The depth of field is obviously the same in both cases. (a) A 12-inch lens at *f*/16, contact print. (b) A 6-inch lens at *f*/8, enlarged 2× in printing.

Focusing a Camera to Cover a Specified Range of Object Distances

This problem is easily solved by eliminating *s* and *h*, respectively, between the two distance formulae (5.7). These elimination processes give

$$s = \frac{2L_1L_2}{(L_1 + L_2)} \quad \text{and} \quad h = \frac{2L_1L_2}{(L_1 - L_2)} \quad . \tag{5.9}$$

Therefore, if we wish to cover the range of distances from $L_1 = 20$ feet to $L_2 = 12$ feet, we must focus our camera on the distance

$$s = \frac{2 \times 12 \times 20}{20 + 12} = \frac{480}{32} = 15 \text{ feet} \quad ,$$

and we must stop our lens down until its hyperfocal distance is given by

$$h = \frac{2 \times 12 \times 20}{20 - 12} = \frac{480}{8} = 60 \text{ feet} \quad .$$

To do this, the entrance pupil diameter must be equal to $h/1500$, or 0.48 inch. This corresponds to an aperture of $f/4$ for a 2-inch lens, or $f/8$ for a 4-inch lens.

(c) Depth of Field for a Fixed Circle of Confusion in the Image

In both of the previous methods of depth calculation we have assumed a knowledge of the manner in which the final image will be viewed, and neither method is convenient for the calculation of depth-of-field tables or scales to be engraved on a lens mount. The manufacturer of the Leica camera, who pioneered the use of 24 × 36 mm images on 35mm film, adopted the assumption that a suitable limiting size of the circle of confusion on the film should be about 1/30 mm or 1/750 inch. As the focal length of the normal lens used on the Leica is 50 mm, this implies that the average observer can just distinguish between a point and a circle of confusion subtending an angle of 1 in 1500, which is the angle we have assumed previously.

To develop depth formulae using the image circle c', we turn Fig. 5.2 around so that it refers to the situation in the image space rather than in the object space (Fig. 5.5). We thus find that

$$L_1' = \frac{ds'}{d + c'} \quad \text{and} \quad L_2' = \frac{ds'}{d - c'} \quad . \tag{5.10}$$

We can now transfer these distances into their images in the object space by use of the following well-known general relationship: if a and a' are the distances of object and image, respectively, assuming both are positive quantities, then $1/a + 1/a' = 1/f$. Hence,

$$L_1 = \frac{f}{1 - \left(\dfrac{f}{L_1'}\right)} \quad \text{and} \quad s' = \frac{f}{1 - \left(\dfrac{f}{s}\right)} \quad . \tag{5.11}$$

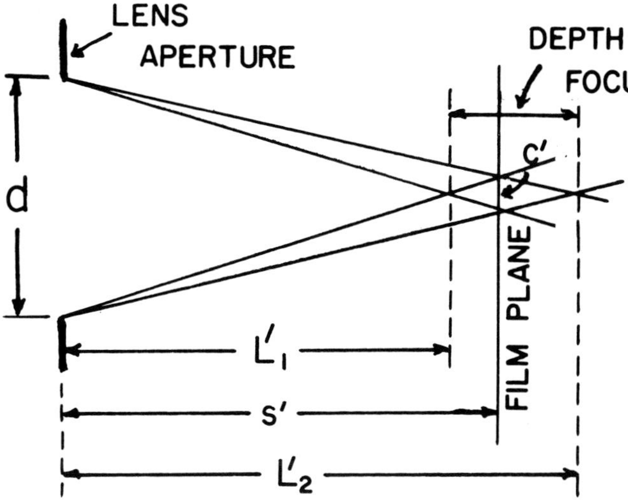

Figure 5.5. Situation in the image space.

Therefore,

$$L_1 = \frac{s}{1 - \frac{c'}{fd}(s-f)} \quad \text{and} \quad L_2 = \frac{s}{1 + \frac{c'}{fd}(s-f)} \quad . \tag{5.12}$$

By the very nature of depth calculations, and the various assumptions that we have made, we are fully justified in replacing the $(s - f)$ in these formulae by the simple s. This has the effect of increasing the calculated depth by an insignificant amount, so that in practice we can write, approximately,

$$L_1 = \frac{s}{1 - \left(\frac{c's}{fd}\right)} \quad \text{and} \quad L_2 = \frac{s}{1 + \left(\frac{c's}{fd}\right)} \quad . \tag{5.13}$$

It is usually more convenient to use the F-number N of the lens rather than the clear aperture, where $N = f/d$. Substituting this into the last equations gives

$$L_1 = \frac{s}{1 - \left(\frac{c'sN}{f^2}\right)} \quad \text{and} \quad L_2 = \frac{s}{1 + \left(\frac{c'sN}{f^2}\right)} \quad . \tag{5.14}$$

The actual depth formulae become

$$D_1 = L_1 - s = \frac{s}{\left(\dfrac{f^2}{c'sN}\right) - 1} \quad \text{and} \quad D_2 = s - L_2 = \frac{s}{\left(\dfrac{f^2}{c'sN}\right) + 1} \ . \qquad (5.15)$$

To determine the hyperfocal distance h, we see that L_1 and D_1 will become infinite if

$$c'hN = f^2 \quad \text{or} \quad h = \frac{f^2}{c'N} \ . \qquad (5.16)$$

Suggested Values of c'

Of course, the recommended size of the acceptable circle of confusion on the film depends on the size of the negative image. It may well be assumed to be equal to, say, 1/1500 of the normal lens focal length, but, as has been remarked, depth of field calculations are highly indefinite and other sizes of the acceptable circle of confusion can be used without causing difficulties. The following table gives sizes of c' that are commonly assumed.*

Film size	**Suggested value of c'**	
	inches	**mm**
8mm cine	1/2000	0.013
16mm cine	1/1500	0.017
110 film	1/1000	0.025
24 × 36 mm	1/750	0.033
$2^1/_4 \times 3^1/_4$ inches	1/250	0.100
4 × 5 inches and up	1/100	0.250

A Fixed-Focus Camera

The concept of hyperfocal distance provides an answer to the question, when is it necessary to focus a camera and when can a fixed-focus camera be used? For example, if we never need to photograph objects closer than 5 feet, and if the hyperfocal distance is 10 feet, it will be unnecessary to adjust the focus. For an 8mm cine camera, for example, using a fixed-focus

*It is of interest to note that in some of Taylor-Hobson's movie camera lenses, the assumed value of c' for 8mm film was taken to be 1/1500 inch, while for 16mm lenses it was 1/750 inch.

lens of 13mm focal length, the effective lens aperture would have to be about *f*/4, assuming a circle of confusion of 0.013 mm for this situation.

In the following table, the values of half the hyperfocal distance for an 8mm movie camera are given for a variety of situations, again assuming a circle of confusion equal to 1/2000 inch, or 0.013 mm:

f	N = 1.9	2.8	4	5.6	8	11
6.5 mm	2.8 ft	1.9 ft	1.3 ft	11.3 in	7.9 in	5.8 in
9 mm	5.3 ft	3.6 ft	2.5 ft	1.8 ft	1.3 ft	11.0 in
13 mm	11.1 ft	7.5 ft	5.3 ft	3.8 ft	2.6 ft	1.9 ft
24 mm	38 ft	25 ft	18 ft	13 ft	9 ft	6.5 ft

These figures represent the nearest object distance, in feet or inches, for which no focusing would be needed on an 8mm movie camera.* For our supposed case in which the closest object is to be at 5 feet, the relation between focal length and *F*-number would be

Focal length (mm)	6.5	9	13	24
F-number	1.1	2.0	4.2	14.4 .

Cameras with Interchangeable Lenses

Two possible situations arise here. In one case the photographer may hold the camera in a fixed location but change the focal length of the lens to stress some important object in the scene. Then, if the final picture is viewed from a fixed distance, the depth of field will fall under case (c) involving a constant circle of confusion c' on the film. On the other hand, if the viewer were to change his or her distance from the picture in order that the particular object of interest would always appear at the same size to the viewer's eyes, at a constant angular subtense, then the situation falls under case (b) and the depth of field becomes much less dependent on the choice of lens.

As an example, suppose that a 16mm movie camera is located 10 feet from the object of interest and that the photographer is using two lenses,

*On one of Kodak's fixed-focus 6.5mm Cine Ektanon lenses for an 8mm camera, the nearest distance of acceptable imagery was coupled to the diaphragm control, stating that

F-number	1.9	2.8	4.0	5.6	8.0	11.0	16.0
Closest object (in feet)	3	2.5	2	1.5	1.25	1	0.75

For this sequence to be valid, the circle of confusion c' in the image had to drop steadily from 0.012 mm at *f*/1.9 to 0.006 mm at *f*/11.

a 1-inch $f/4$ and a 3-inch $f/2.8$. Then, in the first assumed situation, the observer will be at a fixed distance from the projected picture, and the depth of field would be

1-inch lens:	$L_1 = 11.3$ ft,	$L_2 = 9.0$ ft,	total depth = 2.25 ft
3-inch lens:	$L_1 = 10.3$ ft,	$L_2 = 9.8$ ft,	total depth = 0.50 ft

We see that the total depth of field, given by $L_1 - L_2 = 2s^2c'N/f^2$ (approx.), depends only on the factor N/f^2 when c' and s are fixed. For several normal 16mm camera lenses, the value of this factor is

Focal length	**F-number**	**Factor N/f^2**
1-inch	1.4	1.40
2-inch	2.0	0.50
4-inch	2.7	0.17
6-inch	4.0	0.11

Thus, to an observer seated in a fixed position relative to the projection screen, the 6-inch lens used at full aperture ($f/4$) will have about one-twelfth as much depth of field as the 1-inch lens, also used at full aperture ($f/1.4$), when both lenses are used on the same object at the same distance from the camera. If each picture were observed from its correct center of perspective, however, the depth for the 6-inch lens would be about one-half as much as for the 1-inch lens. These results bear out the well-known fact that a long-focused lens on a 35mm or cine camera must be set to the correct object distance with extreme precision, because the depth is so very small.

The great reduction in the depth of field that results from the use of a long-focus lens became painfully evident with the introduction of the various forms of large-format professional motion pictures in the mid-1950s, which of course required lenses of longer focal length than usual.* Since it was essential to maintain the depth of field in the photographs, the relative aperture of the lenses had to be reduced, and hence the illumination on the set had to be increased by a factor that actually worked out to be approximately equal to the square of the ratio of the film dimensions.

Depth of Focus

Care should be taken to distinguish between the *depth of field* in the object space, described above, and what is known as the *depth of focus* inside the

*R. N. Wolfe and F. H. Perrin, "Depth of field and perspective considerations in wide-screen cinematography," J.S.M.P.T.E. **65**, 37-42 (1956).

camera. This latter quantity is a measure of how accurately a camera must be focused so that the tolerable circle of confusion is not exceeded. The situation is indicated in Fig. 5.5. Here, c' represents the diameter of the limiting circle of confusion, and D is the half-value of the depth of focus. Clearly, c' and D are related by the size of the effective F-number of the lens, which is equal to the ratio of the image distance s' to the linear lens aperture diameter d. Hence,

$$\text{Depth of focus} = \pm\left(\frac{s'}{d}\right)c' \ . \tag{5.17}$$

This quantity is surprisingly small. For example, if a 2-inch $f/2.8$ lens is focused on an object 10 feet away, the image distance s' is 2.03 inches, and the aperture d is 0.714 inch, so that if we can assume a value of 1/750 inch for c', the depth of focus becomes ±0.0038 inch, or about ±0.1 mm. Everything in the camera connected with the focusing adjustment must be held to this order of precision.

It should be noted that the depth of field and the depth of focus are related by the ordinary longitudinal magnification, which is equal to the square of the transverse magnification when the longitudinal displacement is small (see page 34).

In many cameras the depth-of-field indicators are engraved alongside the distance pointer on the focusing scale of the lens (Fig. 5.6). The locations of these marks correspond to a lens movement equal to the depth

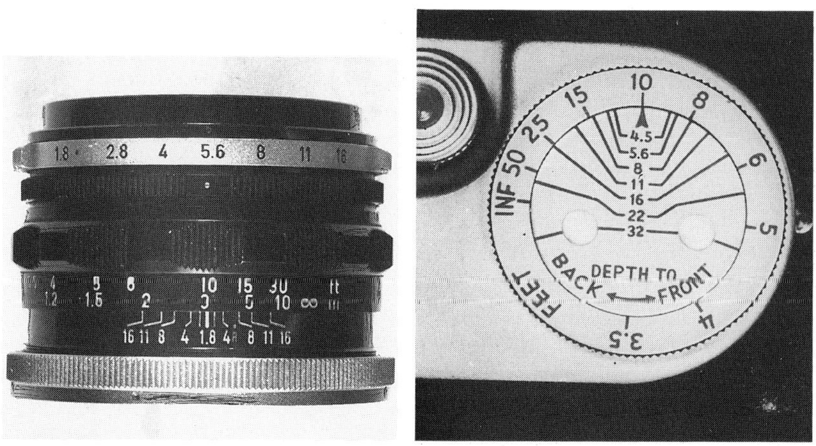

(a) (b)

Figure 5.6. Typical depth-of-field scales. (a)On a camera lens barrel. (b) An auxiliary depth indicator.

of focus at each stated *F*-number, and their positions can be immediately located once the pitch of the focusing thread is known. In some cameras the depth-of-field dial was entirely separate from the lenses and had to be set manually.

There is, of course, some degree of approximation here, as the image distance varies when the camera is focused on a close object, and the far depth is generally longer than the near depth, but these approximations are considered unimportant in view of the indefinite nature of depth of field.

In the case of a zoom lens, the image distance from the second principal point to the focal plane varies considerably, as it is equal to the focal length,

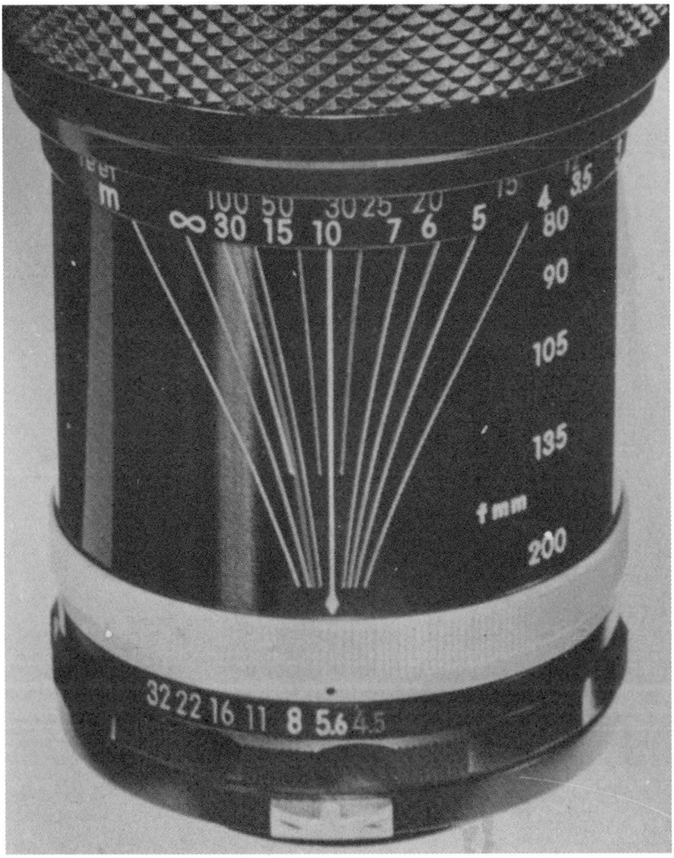

Figure 5.7. A depth-of-field scale for a 35mm zoom lens, covering a focal-length range from 80 to 200 mm at *f*/4.5.

so the depth of focus varies with the zoom setting. In some lenses this situation has been taken care of by marking a set of curved depth limit lines on the lens barrel, which are uncovered by the sliding zoom control sleeve on the lens (Fig. 5.7).

Depth of Field at Fixed Magnification

The question often arises, especially when making almost life-size photographs of insects or small objects, whether we shall obtain more depth by using a lens of long or short focal length, supposing that the same *F*-number *N*, and magnification *m*, are used in both cases.

A mathematical analysis of this problem is almost unnecessary, since in both cases the effective *F*-number of the imaging cone is $N(1 + m)$, where *N* is the *F*-number of the lens and *m* is the magnification (see page 108). The depth of focus in the image in both cases is therefore $\pm\, c'N(1 + m)$, no matter what may be the focal length of the lens. The depth of field (in the object) is in both cases equal to

$$\pm \frac{c'N(1 + m)}{m^2} = cN\left(1 + \frac{1}{m}\right),$$

if *c* is the acceptable circle of confusion in the object. This result has an important application wherever a fixed ultimate magnification is required, such as in motion pictures, portraiture, photomicrography, and television. The *F*-number is the only thing that counts, so far as depth is concerned, and any convenient focal length can be used.

In Fig. 5.8, four photographs of an inclined ruled test chart are shown, which were made using lenses of focal length ½, 1, 2, and 4 inches respectively. All four photographs were made at 10× magnification and *f*/4.5 aperture, and they all exhibit identical depth of field.

This rule does *not* apply if the focal length is short and the lens diameter is very small, because then the hyperfocal distance becomes small enough to be comparable with the object distance. In that case, the far depth becomes very large and may even become infinite with a short-focus lens, whereas it will remain at a finite value with a long-focus lens.

Depth of Field for a Sloping Object

If we have a camera that lacks a swing back, we must stop the lens down sufficiently to ensure that the depth of field will include all parts of the object. However, by tilting either the film plane or the lens so that the median plane of the sloping object is properly imaged on the film, the

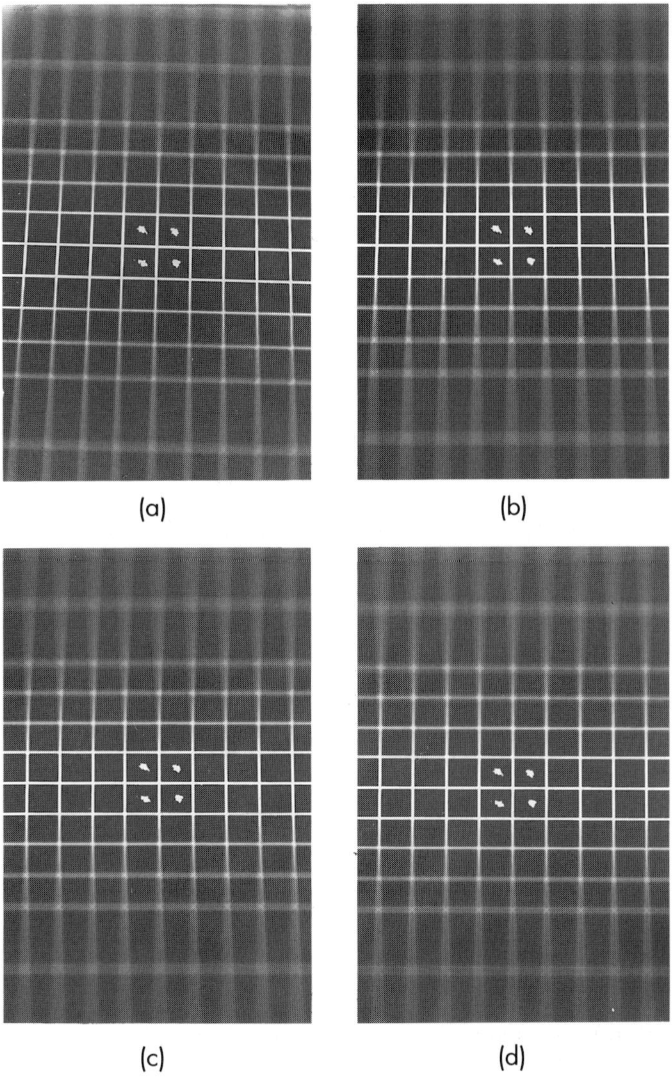

(a) (b) (c) (d)

Figure 5.8. Photographs of a sloping ruled test chart, taken at a magnification of 10× and aperture *f*/4.5 with lenses of four different focal lengths: (a) 13 mm, (b) 25 mm, (c) 50 mm, and (d) 101 mm.

useful depth of field will be very much increased and larger lens apertures may be employed (Fig. 5.9). There will be some keystone distortion resulting from the use of a swing back, but this can be rectified afterward in printing. Care should, of course, be taken to see that none of the desired

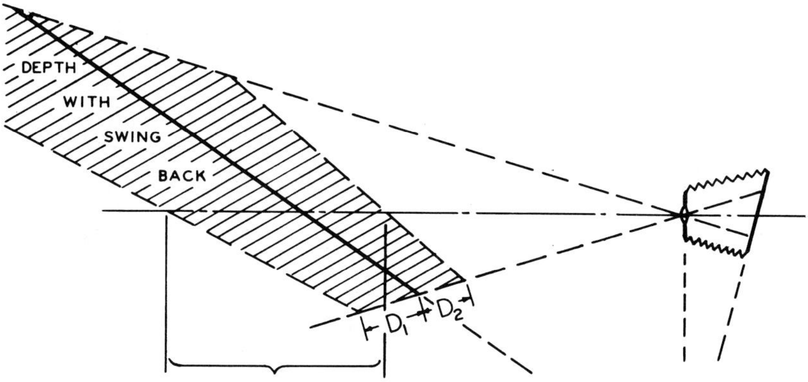

Figure 5.9. Depth of field using the swing back.

objects falls outside of the tilted area of acceptable definition indicated in the diagram.

Most modern SLR cameras are equipped with a "depth preview" button, which stops the lens down to its ultimate value as determined by the exposure control mechanism so that the photographer can see in advance what depth will exist in the picture. This is assuming that the final picture will be viewed from a distance corresponding to the viewing distance in the camera viewfinder.

Effect of Finite Resolving Power of the Emulsion

Since the size of the acceptable circle of confusion in the image c' is approximately equal to $f/750$, this quantity represents the limit of barely acceptable resolution in the negative. For a film and lens combination having a resolving power of 60 line pairs per millimeter, the film resolution will only begin to affect depth of field when the focal length is less than about 750/60 = 12.5 mm. Thus, ordinary emulsions play a negligible part in determining depth of field except when very small lenses are used, such as those on 8 mm movie cameras, or when the emulsion of the film or the print is so grainy that the usual resolving-power figure of 60 line pairs per millimeter can no longer be applied.

Effect of Lens Aberrations

If the lens aberrations are great enough to produce an aberrational "star image" larger than one-thousandth of the focal length of the lens, then our simple depth-of-field theory will be upset because the aberrations of the lens will be visible to the observer as a deterioration in the sharpness of the image. Spherical aberration, chromatic aberration, field curvature, and astigmatism all cause a longitudinal displacement of the best image from the paraxial focal plane, and thus the depth of field may be aided for objects situated closer than the focused plane and reduced for objects lying beyond the focused plane, or vice versa. For instance, many lenses have a noticeable amount of undercorrected spherical or zonal aberration, which tends to form an acceptable image of objects lying considerably *closer* than the focused plane, but it does not in any way aid the definition of more distant objects. This has the effect of making the near-depth equal to or even greater than the far-depth, whereas according to elementary theory the near-depth must always be less than the far-depth. In some "soft-focus" lenses the residual spherical and chromatic aberration has been deliberately made so large that there is no definite focal plane at all, and objects lying at a very wide range of distances from the lens all appear equally sharp. This is the basis of many lenses and lens attachments that claim to give increased depth of field.

For this reason, and because the apparent size of near objects is larger than that of distant objects, a depth-of-field table or scale should be taken with many reservations, and depths stated to small fractions of an inch can be somewhat misleading. Moreover, the observer's eye is not always situated at the correct center of perspective, and hence the depth may be multiplied or divided by a factor depending on the observer's departure from the proper viewing conditions, which may reach as much as 2 or 3 or more. Thus, hair-splitting arguments about depth-of-field data become inconsequential.

Conclusions on Depth of Field

Information of practical value from the discussion given in this chapter may be summarized as follows:

(1) The smaller the stop, the greater the depth.

(2) The depth decreases rapidly as the subject approaches the camera. It is, therefore, important to determine the distance of near subjects more carefully than of distant subjects. (It is not possible for the average person to guess a six-foot distance accurately enough if working at *f*/4.5.)

(3) For the same magnification on the film (not the same subject distance) and the same *F*-number, lenses of all focal lengths have the same depth of field, except when the lens diameter is very small.

(4) For the same subject distance, the depth decreases with increasing focal length. A 35mm camera negative, enlarged to the size of a large format camera contact print, shows greater depth than the latter, provided both are taken at the same distance and relative aperture.

(5) For prints enlarged to equal size, all lenses of the same linear aperture (not relative aperture) have the same depth. A 16mm cine camera with a 25-mm lens at *f*/1.9 has the same depth as a large camera with a 135-mm lens at *f*/11; at these settings, the effective linear aperture of both lenses is 12.5 mm.

(6) Depth can be greatly increased by tilting the lens so that the plane of the film, the plane of the lens, and the average plane through the subject all meet at a common point.

Chapter 6

The Brightness of Images

The relation between the aperture of a lens and the brightness of the image produced by it on the photographic emulsion is often misunderstood, yet it is of the greatest importance to the photographer who wishes to make the best use of the equipment. The tremendous efforts of lens designers and manufacturers that have been devoted to the production of lenses of extremely high relative aperture are an indication of the need that exists for brighter images and "faster" lenses.

In this chapter, we are concerned with the flow of light from an object, through a lens, to the image. Several photometric terms must be understood before we can give a precise statement of this effect, and of the factors that control the brightness of the image projected on the film in a camera.

The *illumination* (illuminance) produced by a lamp at any distance from it is found by dividing the candle power of the lamp by the square of the distance (the inverse square law). Thus, a 50-candle lamp will produce, at a distance of 3 feet, an illumination of 50/9 = 5.6 foot-candles. The illumination in a well-lighted factory or classroom may reach 50 foot-candles, and in motion-picture or television studios, illuminations as high as 200 to 300 foot-candles are common.

The term *flux* is used to express a quantity of light. The unit of flux is the lumen, defined as the amount of light falling on each square foot of a surface under an illumination of 1 foot-candle; hence, foot-candles and lumens per square foot are two ways of expressing the same thing. The convenience of this term may be seen by an example. Suppose we know that a certain 16mm projector emits 550 lumens. Then, if the projected image is 3 × 4 ft, the average illumination on the screen will be 550/12 = 46 foot-candles; if the image is 5 × 6.6 ft, the illumination will be 550/(5 × 6.6) = 16.7 foot-candles, and so on.

The *brightness* (luminance) of an object is expressed by its candle power per unit area. Thus, the filament of a biplane 750-watt projection lamp has about 2500 candle power and an area of 10×10 mm; its brightness is thus about 25 candles per square millimeter. This is quite a high value of brightness, but it is exceeded by the arc lamps used in motion-picture projection, which sometimes reach several hundred candles per square millimeter, and by the surface of the sun, which has a brightness of over 2000 candles per square millimeter.

At the other end of the scale the brightness of a piece of white paper under ordinary room lighting, for example, is quite low. The formula for calculating the brightness of an illuminated object is

$$B = \frac{kE}{\pi} , \tag{6.1}$$

where k is the reflectivity of the surface, say 0.9 for white paper, E is the illumination in foot-candles falling upon the paper, and $\pi = 3.14$. Thus, if the illumination in the room is moderately high, say 50 foot-candles, the brightness of white paper will be $0.9 \times 50/31.4 = 14$ candles per square foot. This is equal to $14/930 = 0.015$ candle per square millimeter, since there are 930 square millimeters in 1 square foot.

The formula (6.1) connecting illumination with the brightness of a surface is not always applicable. For example, a *specular* surface such as sandblasted metal, metallic paint, or a ribbed or beaded projection screen, tends to reflect light somewhat like a mirror, with the angle of reflection equal to the angle of incidence. Such a surface will therefore appear to be brighter than white paper in the direction of the specular reflection, and duller than white paper in other directions. This subject is discussed later in this chapter (page 135).

The inconvenience of the factor π in this formula (6.1) is so great that a new brightness unit has been developed called the foot-lambert, of size equal to $1/\pi = 0.32$ candles per square foot. Hence, the number of foot-lamberts required to express the brightness of a surface is equal to π times the number of candles per square foot. In terms of this new unit, Eq. (6.1) becomes simply

$$B_L = kE \text{ foot-lamberts} . \tag{6.2}$$

We conclude, therefore, that *the brightness in foot-lamberts of a perfectly reflective and perfectly diffusing surface is just equal to the illumination in foot-candles falling upon it.*

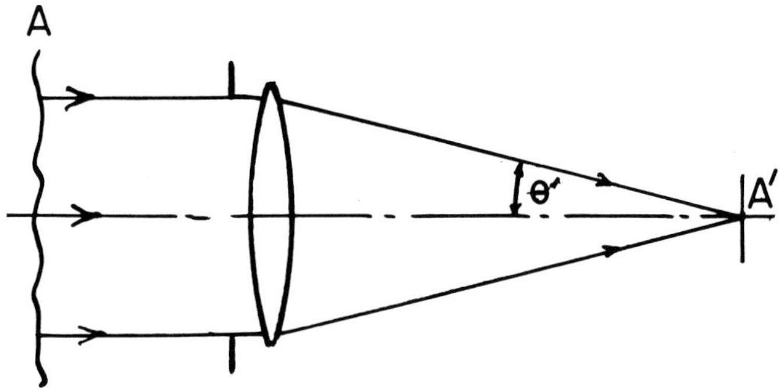

Figure 6.1. The illumination at a point on the lens axis.

The Illumination in an Optical Image

This is the major question to be considered in this chapter, as it underlies all the brightness problems that are likely to be encountered in projection, enlarging, or ordinary photography through a lens.

The situation is indicated in Fig. 6.1. An extended uniform diffusing source A fills the lens aperture with light. A fraction t of this light is then radiated by the lens into the axial point of the image plane at A′, where it produces an illumination E. The relation between the luminance B of the extended source and the image illumination E can be shown to be

$$E = \pi \, t \, B \sin^2 \theta' \ , \tag{6.3}$$

where θ' is the half-angle of the cone of light proceeding from the lens to the image. It should be especially noted that the image illumination depends only on the luminance of the source, the transmittance* of the lens, and the angle θ' of the imaging cone. This is a very important result, and it provides the basis for the whole F-system of lens-aperture markings.

If the object brightness B_L is expressed in foot-lamberts, the formula becomes

$$E = t \, B_L \sin^2 \theta' \ . \tag{6.4}$$

*The transmittance t of a lens is defined as the ratio of the emerging flux to the entering flux contained in a beam of light so narrow that it is not limited by any mechanical obstructions on its passage through the lens.

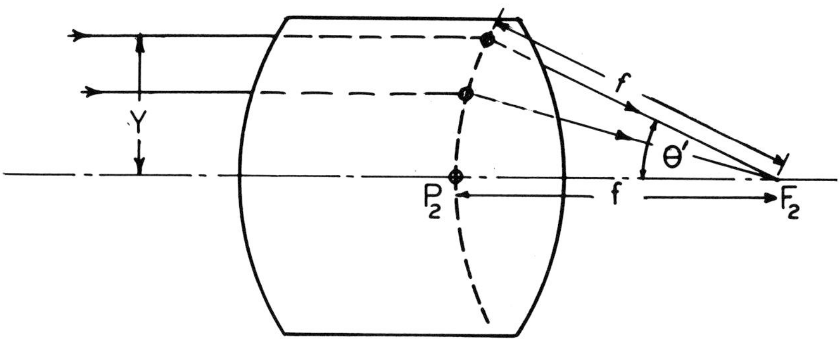

Figure 6.2. The equivalent refracting surface of a well-corrected lens.

The Relative Aperture of a Lens

If a lens having a focal length f and a circular aperture of semidiameter Y is used to photograph a distant object, the situation is shown in Fig. 6.2. The term sin θ' in Eq. (6.3) may be replaced* by the ratio Y/f, and the illumination equation becomes

$$E = \frac{t\,\pi\,B\,Y^2}{f^2} \,. \tag{6.5}$$

The quantities Y and f in this formula may now be replaced by the familiar F-number of the lens, N, which is defined as the ratio of the focal length to the aperture diameter. Hence,

$$F\text{-number} = N = \frac{f}{2Y} = \frac{1}{2\sin\theta'} \,, \tag{6.6}$$

and

$$E = \frac{t\,\pi\,B}{4N^2} \,. \tag{6.7}$$

* It is a common error to suppose that the ratio of Y/f is actually equal to tan θ', and not sinθ' as stated in the text. The tangent would, of course, be correct if the principal planes were really plane. However, the complete theory of the Abbe sine condition shows that if a lens is corrected for coma and spherical aberration, as all good photographic objectives must be, the second principal plane becomes a portion of a sphere of radius f centered about the focal point, as is correctly shown in Fig. 6.2. In this case, Y/f is equal to sinθ', and it is evidently impossible for any lens corrected for those two aberrations to have an aperture greater than twice the focal length. Thus, no well corrected photographic lens can have a relative aperture greater than f/0.5.

The required exposure to take photographs of distant objects is, therefore, *directly proportional to the square of the F-number, and independent of the actual focal length of the lens*. It is surprising, at first, to learn that a 12-inch *f*/4 lens with a 3-inch diameter aperture will form just as bright an image on the film as a small 1/2-inch *f*/4 cine lens with a linear aperture of only 1/8 inch. The explanation is that the area of the picture formed by the larger lens is $(24)^2 = 576$ times as great as the area of the picture formed by the smaller lens, the areas of the two lens apertures being in exactly the same proportion.

In some European cameras the relative aperture of the lens is expressed as a pure ratio, for instance, 1:4.5; but it is the usual British and American custom to state the aperture as a faction of the focal length, e.g., *f*/4.5. The F-number, 4.5, has the same meaning in the two cases.

A series of some typical lens apertures, with the corresponding relative lens speeds based on a maximum of 1000, is shown in the following table:

F-number, N	0.5	1	1.5	1.9	2.5	3.5	4.5	6.3	8	11
Angle θ'	90°	30°	19.5°	15°	11.5°	8°	6.5°	4.5°	3.5°	2.5°
Relative speed	1000	250	111	69	40	20	12	6.3	3.9	2.1

Effective Lens Aperture at Finite Magnification

When a reasonably thin lens is used with a near object, at a magnification m, such as in a copying camera or an enlarger, the image distance becomes $f(1 + m)$, and the effective F-number of the lens is equal to its true F-number multiplied by $(1 + m)$. The formula for image illumination then becomes*

$$E = \frac{\pi t B}{4(\text{F-number})^2 (1 + m)^2} \quad . \tag{6.8}$$

This is a most useful rule to remember, and it is referred to several times in this book.

* Strictly speaking, the second bracketed term in the denominator should be $(1 + m/m_p)$, where m_p is the pupil magnification, i.e., the diameter of the diaphragm as seen from the rear divided by its diameter as seen from the front of the lens. In a perfectly symmetrical lens, $m_p = 1$, and it is sufficiently close to 1.0 in most normal types of lens. However, in a telephoto lens m_p is likely to be less than 1.0 and in a reversed telephoto lens it is greater than 1.0. Actually, this problem seldom arises because neither of these lens types should be used with very close object distances.

The T-Stop System

Because the F-number of a lens as defined in Eq. (6.6) takes no account of the lens transmittance t, a move has been pioneered by the motion-picture industry* to replace the F-number by the so-called T number, defined in the following way:

$$T\text{ number} = \frac{F\text{-number}}{\sqrt{t}} \quad . \tag{6.9}$$

Thus, the image illumination formula (6.7) becomes, in terms of T numbers,

$$E = \frac{\pi B}{4\,(T\text{ number})^2} \quad .$$

All lenses at the same T number will give the same axial illumination within the image of the same object, independent of the lens transmittance t. Before lens coating became general, transmittances could vary from 0.9 for a simple box-camera lens down to as low as 0.5 for a complex anastigmat. Hence, if these two lenses were both set at $f/11$, for instance, their T stops would be T-11.5 and T-15.5, respectively, and the image illumination produced by the simpler lens would be 1.8 times as bright as that produced by the complex lens. Although this difference would not be too serious in everyday black-and-white photography, it could be serious in a motion-picture studio where strips of film of the same subject, made in rapid succession by several cameras, would be spliced together for projection. In color cinematography especially, where small degrees of overexposure or underexposure are readily noticeable, the demand for a photometric type of lens aperture calibration was strongly felt. Now that all lenses are antireflection coated, the need for a T-stop calibration has almost disappeared. Nevertheless, some motion-picture camera lenses are still being marked with T stops, and of course T stops are essential in any equipment in which a beamsplitter is inserted in front of a camera so that only part of the light reaches the film.

Fraction of a Stop

We often hear photographers referring to "half a stop" in connection with variations in lens aperture or exposure time. Since exposure varies with the

*American Standard PH 22.90-1953.

square of the *F*-number, i.e., with the area of the lens aperture, a whole stop corresponds to a 2× or 0.5× change in the lens area, or a $\sqrt{2}$ or $\sqrt{0.5}$ change in the stop diameter. A half stop therefore represents a $\sqrt[4]{2}$ increase or decrease in the stop diameter, and so on. The following table may be useful in this regard:

Fraction of a stop	Proportional change in stop diameter		
Whole stop	$\sqrt[2]{2}$ = 1.414	or	0.707 (41% up or down)
Half stop	$\sqrt[4]{2}$ = 1.189	or	0.841 (19%)
One-third stop	$\sqrt[6]{2}$ = 1.122	or	0.891 (12%)
One-quarter stop	$\sqrt[8]{2}$ = 1.091	or	0.917 (9%)
One-sixth stop	$\sqrt[10]{2}$ = 1.059	or	0.944 (6%)

The Speed of an Emulsion

During the past century, numerous attempts have been made to determine the speed of an emulsion, a factor obviously of the greatest importance to the photographer. Two basic procedures have been followed: The first is to determine by trial the minimum exposure of an average outdoor scene from which an acceptable print can be made (for black-and-white photography) or which will produce an excellent color slide (for Kodachrome or other reversal systems). The second procedure is to determine in the laboratory some specific property of an emulsion, which would hopefully correlate with the working speed of the emulsion in actual practice.

The first approach was used by several of the makers of early exposure meters, such as Watkins and Wynne, and the makers of the first photoelectric exposure meters, Weston and G.E.C. These companies made their own determinations of the speed of the emulsions currently available in the open market and published lists of these data for sale alongside their instruments. The second approach was used in the Scheiner and early DIN systems, where the exposure required to produce a just detectable, or just measurable, blackening of the emulsion was found. A similar but more elaborate method was used by Hurter and Driffield in England in establishing their H&D system, involving the plotting of a characteristic curve of the emulsion in which density is plotted as ordinate against the log of the exposure (illumination times time) as abscissa.

It is obvious that, no matter what procedure is used, it is essential that the type of illumination and the time and temperature of the development must be carefully controlled, and also used by the photographer, if emulsion speed data are to have any meaning. It is interesting to note that up to World War II the Kodak Company did not publish any emulsion speeds for their products, their argument being that the whole matter was too uncertain to be readily quantized. Finally, during the war, the American Standards Association (ASA), now known as the American National Standards Institute, adopted a procedure based on the first method described above, and established the so-called "exposure equation":

$$T = \frac{N^2}{BSt} , \tag{6.10}$$

where T is the minimum exposure time in seconds required to produce a satisfactory negative or color slide, N is the F-number of the lens used on a camera, S is the ASA emulsion speed, and B is the average scene brightness measured in candles per square foot. The emulsion speed is then determined by making a series of exposures on the test film using carefully controlled values of T, B, and N, whence the speed is found by

$$S = \frac{N^2}{TB} .$$

(The last factor, t, representing the lens transmittance, is so close to unity that it is generally ignored.)

It is interesting to note that for most ordinary films used by the average photographer, the ASA speed came out to be about midway between the well-established Weston and G.E.C. speeds. The ASA system has gradually achieved international acceptance, and it is now officially adopted by the International Standards Organization as the ISO speed.

The current logarithmic DIN speeds are defined by

$$\text{DIN} = 10 \log \text{ASA} + 1 . \tag{6.11}$$

A helpful rule for estimating exposure time is the following: *At f/16 in bright sunlight, the average landscape requires an exposure time in seconds equal to the reciprocal of the ASA film speed.* The justification for this rule can be seen from Eq. (6.10), because the brightness of such a scene is about 250 candles per square foot, the square root of 250 being about 16.

It is, of course, impossible to develop a precise conversion table between the various speed systems, as they are all based on different criteria. However, there are a few approximate relationships that can be sometimes useful, but they must not be regarded as accurate equivalents. Roughly, then, we may say that

$$\text{Scheiner} = \text{about 7 times log (H \& D)}$$

$$\text{DIN} = \text{Scheiner minus 8}$$

$$\text{ASA} = \text{antilog}\left(\frac{\text{Scheiner} - 9}{10}\right) .$$

Hence, if the speed of an early film was 700 H & D, the Scheiner speed would be about 20, the DIN speed about 12, and the ASA speed about $12^1/_2$.*

The Additive Photographic Exposure System (APEX)

This is a highly logical logarithmic form of the basic exposure equation and was developed in Germany about 1954. Unfortunately, it never became popular in this country in spite of its many advantages. It is defined in American Standards PH-2.5 and PH-2.12.

The system is based on four "values," namely, the *aperture* value A_v of the diaphragm opening, the *time* value T_v of the shutter, the *speed* value S_v of the film, and the *brightness* value B_v of the subject. Then the *exposure* value E_v is related to these values in the following way:

$$E_v = A_v + T_v = S_v + B_v \ . \tag{6.12}$$

The aperture and time values are determined by the camera settings, while the speed and brightness values are determined by the film and subject conditions.

These four values are actually powers of 2, namely,

$2^{A_v} = N^2$. Thus, $A_v = 6.64 \log N$, where N is the F-number of the lens.

$2^{T_v} = 1/T$. $T_v = -3.32 \log T$, where T is the exposure time in seconds.

$2^{S_v} = S/\pi$. $S_v = 3.32 \log S - 1.66$, where S is the ASA film speed.

$2^{B_v} = B_L = \pi B$. $B_v = 3.32 \log B + 1.66$, where B is the scene brightness in candles per square foot and B_L

*An excellent survey of the problems involved in sensitometry is given in the *Handbook of Photography* by Henney and Dudley, Chapter 7, McGraw-Hill, New York (1939). Other good references are *Photography for the Scientist*, C. E. Engel, ed., Academic Press Inc., New York and London (1968); *SPSE Handbook of Photographic Science and Engineering*, W. Thomas, ed., Wiley-Interscience, New York and London (1973).

is the scene brightness in foot-lamberts. The factor 3.32 is the reciprocal of log 2. All the logs are to base 10.

The following table will be helpful in determining the four values to be used in any situation. Note that increasing any of the factors by 1.0 has the effect of doubling or halving the exposure.

N (F-number of lens)	A_v	T (Exposure time of shutter)	T_v	S (ASA film speed)	S_v	B (Scene Brightness) fL	cd/ft^2	B_v
1	0	1 sec	0	3.1	0	1	0.3	0
1.4	1	1/2	1	6.2	1	2	0.6	1
2	2	1/4	2	12.5	2	4	1.3	2
2.8	3	1/8	3	25	3	8	2.5	3
4	4	1/15	4	50	4	16	5.1	4
5.6	5	1/30	5	100	5	32	10.2	5
8	6	1/60	6	200	6	64	20.4	6
11.3	7	1/125	7	400	7	128	40.7	7
16	8	1/250	8	800	8	256	81.5	8
22.6	9	1/500	9	1600	9	512	163.	9
32	10	1/1000	10	3200	10	1024	326.	10

In practice, these figures may, of course, be drastically rounded off, especially with black-and-white film.

As an example, suppose that our subject brightness is 64 foot-lamberts and that we are using a film having an ASA speed of 100. Our camera must therefore be set at an exposure value E_v equal to 11, because the S_v of the film is 5 and the B_v of the scene is 6. This exposure value of 11 can obviously be obtained in many ways, for instance,

1 s at *f*/64

1/2 s at *f*/32

1/4 s at *f*/22

down to

1/1000 s at *f*/1.4 .

In some recent cameras the shutter-speed and lens-aperture dials are interlocked so that their combined E_v appears directly on a dial. The latest models of the Weston exposure meter also show the E_v value corresponding to any setting of the computing dials. It is worth noting that the exposure value of most simple box cameras is about 13 (1/40 s at *f*/14 or 1/80 s at *f*/10).

Because it is difficult to visualize scene brightnesses, the following list of suggested values may prove useful:

	B_v
Bright sun on light sand or snow	11
Bright or hazy sun with distinct shadows	10
Weak hazy sun	9
Cloudy bright; sunsets	8
Heavy overcast; open shade	7
Daylighted table near window	6
Two No. 2 photofloods at 8 feet	4
Bounce light, 2 reflecting floods on white ceiling	2½
Sports arenas	2 to 3
Modern light kitchens	1
Four 300-watt floor lamps	0
Night interiors, etc.	–2 to 0
Dusk outdoors 45 min. after sunset	–5 to –4
Full moonlight	–8

Photoelectric Light Meters

During the past fifty years many types of photoelectric light meters have appeared on the market. At first these consisted of a copper-oxide photovoltaic cell connected directly to a micro- or milli-ammeter.* In these instruments the meter reading is roughly proportional to the total light flux falling on the cell, and provided there is no obstruction in front of the cell, the meter can be calibrated to read the incident illumination in footcandles. Such meters are in general use by illuminating engineers for this purpose.

However, for photography a much more useful figure is the average brightness of the scene to be photographed. It can be seen by Eq. (6.3) that if the field of view of the cell is limited to some fixed angle θ', the illumination on the cell will be proportional to the average scene brightness B, and by holding the cell so as to face the scene, the meter reading can be calibrated directly in candles per square foot as required in the exposure equation (6.10). Today many exposure meters are equipped with a CdS cell and a small battery, and the cell is exposed to a restricted field angle θ', with the result that a highly precise indication of the desired exposure can be indicated.

*W. N. Goodwin, Jr., "The photronic photographic exposure meter," J.S.M.P.E. **20,** 95–118 (1933).

In recent years several incident light exposure meters have been manufactured, a typical example being the Norwood Director. This is equipped with a translucent hemispherical "photosphere" over the photocell surface to collect and integrate the light falling on it from various directions, and consequently the scale does not read foot-candles in the usual sense because oblique light has a relatively greater effect than it would have on a flat surface.

The earliest SLR cameras commonly had an external exposure meter mounted on top, which gave a visible measure of the scene brightness for the photographer to use in calculating the exposure time. Later SLR cameras were equipped with through-the-lens (TTL) metering, a far better arrangement from every point of view. The older arrangement merely indicated the scene brightness, whereas in the TTL method the image illumination actually falling on the film is measured, requiring no calculation to determine the exposure time.

In some newer cameras the light reflected from the surface of the film is measured, assuming that the emulsion has a known reflectivity. This type of metering is so rapid that it can be used to control a built-in flash mechanism, so that when enough light has been received the flash is automatically turned off.

Cameras with Matched Pointers

Two methods of manual exposure setting have been used in SLR cameras. In one type the iris ring must be turned manually by the user until a moving pointer in the viewfinder field reaches a fixed fiducial mark. The action depends on a rearrangement of Eq. (6.12), namely,

$$\underset{\text{(aperture)}}{A_v} - \underset{\substack{\text{(scene}\\\text{brightness)}}}{B_v} = \underset{\text{(film)}}{S_v} - \underset{\text{(shutter)}}{T_v} \ .$$

In these cameras the film-speed dial is mounted on top of the shutter-speed control knob, the two dials working in opposite directions to give the subtraction required in the formula. The combined motion causes a rotation of the illumination meter, which therefore needs a different amount of light to move the pointer to the fiducial mark.

In the other type there are two pointers that the user brings into coincidence, one pointer being moved by the combined film-and-shutter control, while the other is moved by the light from the subject. At first the diaphragm had to be actually closed to vary the illumination on the photocell, but later models required only the setting of the aperture dial on the lens, assuming that the diaphragm would be closed to that reading when the exposure is made.

Cameras Equipped with Automatic Exposure Control

In 1938 the Eastman Kodak Company produced a camera with automatic exposure control called the "Super Kodak 620." This camera was equipped with a photovoltaic cell and a microammeter. When the shutter was cocked, the diaphragm was opened fully, and when the release lever was pressed, a clamp first locked the pointer of the microammeter in place, and then an arm connected to the iris diaphragm was moved along by a spring until it came against the pointer, which prevented further closing of the iris and hence controlled the exposure. Manual changes in shutter speed caused a vane to move in front of the photocell and thus indirectly affected the iris opening. For various reasons, this camera was not popular, but since about 1956 a similar mechanism has been applied to other cameras with considerable success. The exposure control used on the Kodak "Starmatic" camera, for example, is shown diagrammatically in Fig. 6.3. Pressure of the finger on the shutter release lever A raises one-half of the square cat-eye diaphragm and lowers the other half until motion is stopped by contact with the needle of the photocell at *B*. As the upper edge of the thin metal plate forming the upper half of the diaphragm is sloping, the ultimate aperture of the diaphragm depends on the lateral position of the needle.

Some recent 35mm cameras have an automatic exposure control based on the discharge of an electrical condenser through a resistance.* The "resistance" is actually a photocell that changes its resistivity when light falls on it. The condenser is charged before the exposure, and when the shutter opens, the condenser begins its discharge through the photoresistive cell. At the point when the voltage has dropped to some preset level, the shutter closes. Often the user has two choices available: In one, called *shutter priority*, the user is able to set the shutter and the automatic mechanism adjusts the diaphragm to give the correct exposure; in the other, called *aperture priority*, the diaphragm is set by the user and the shutter speed is adjusted accordingly. The former option is more common, as it is difficult to construct a shutter that can be set at any desired speed.

*For a clear description of many of these mechanisms, the reader is referred to an article by L. W. Eisener in "Modern Photography," pp. 66–71 (July 1960).

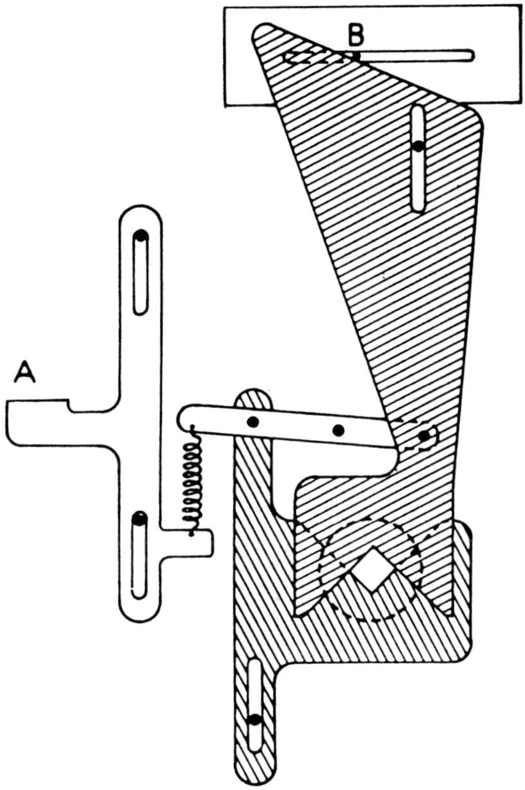

Figure 6.3. Exposure control on the Starmatic camera.

In modern cameras the whole process is controlled by an integrated circuit ("chip") built into the camera.

Programmed Automatic Exposure

In some recent cameras the automatic exposure can be "programmed." In this arrangement both the shutter speed and the lens aperture are varied simultaneously in such a way as to provide the optimum exposure for the particular image brightness. In some cases the program is changed according to the focal length of the lens, because a long-focus lens is harder to hold steady than a short-focus lens, hence a shorter exposure at a wider aperture is needed. Indeed, in some of the most complex cameras, the program is changed as the focal length of a zoom lens is increased.

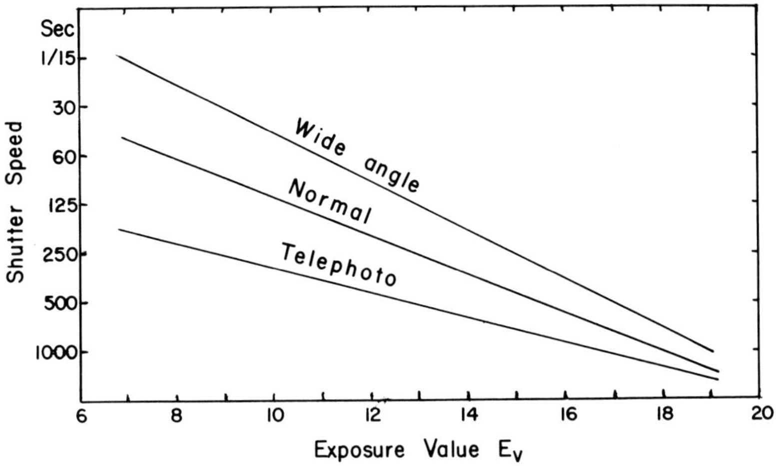

Figure 6.4. A "program" for automatic exposure.

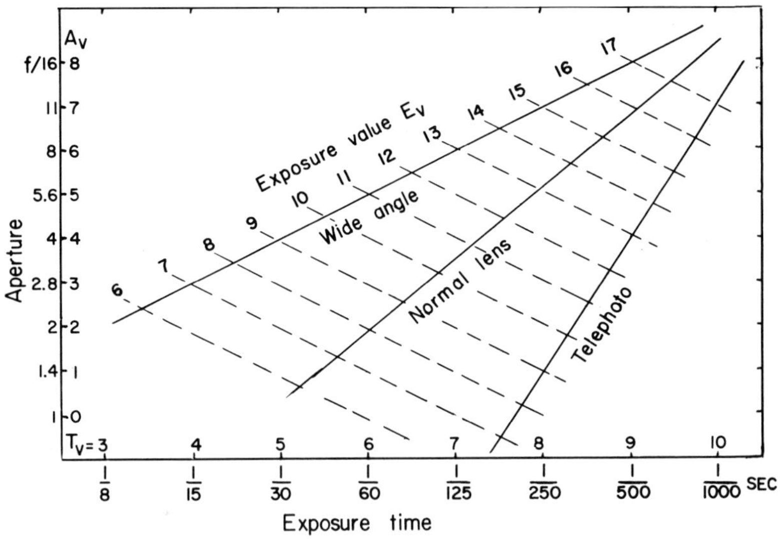

Figure 6.5. Programmed exposure in the Minolta Maxxum 7000 camera.

A typical program is indicated in Fig. 6.4. This graph shows how the shutter speed is varied automatically to suit the image brightness, for three different ranges of focal length. A similar plot is shown in Fig. 6.5, representing the program of the Minolta 7000 camera. It is shown how both the shutter and diaphragm are altered to suit the exposure value E_v of the scene. As was pointed out on page 112, exposure value is a function of both the emulsion speed and the subject brightness.

Automatic Exposure Control in Motion-Picture Cameras

When automatic exposure control is applied to a motion-picture camera, a different mechanism is used because the lens aperture must vary continuously as the light intensity changes, independent of the normal operation of the camera. In this case, the pointer of the microammeter is replaced by a delicately balanced sector in which is cut a spiral slot, opening to the whole lens aperture at one end and closing to perhaps one-tenth of this aperture at the other end (Fig. 6.6). If a single vane is used, the slot-shaped lens aperture that is formed at the low end of the spiral is undesirable, and a pair of "ears" or obstructions are added behind the slot so that the aperture is dumbell shaped at the maximum and is a small rectangle at the minimum aperture. At intermediate points the aperture becomes H-shaped (Fig. 6.7), but this does not seem to be objectionable in practice. The tiny

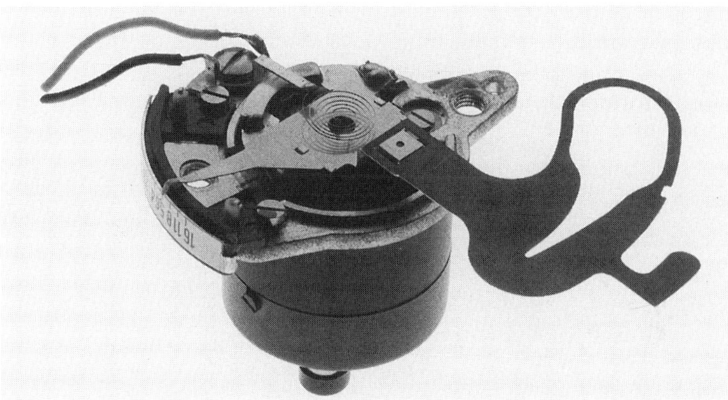

Figure 6.6. Single-vane aperture control used on some 8mm cameras.

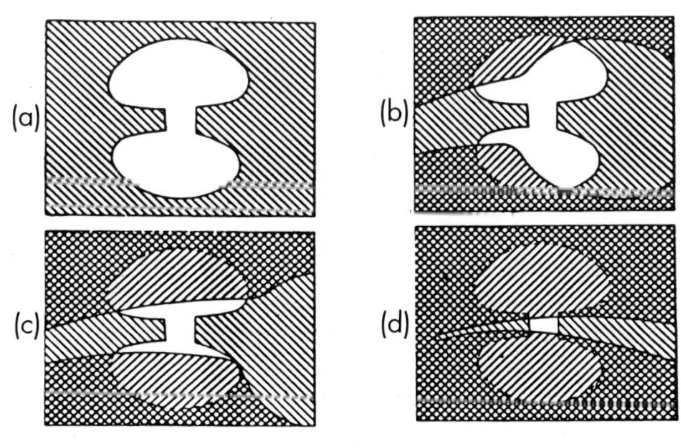

Figure 6.7. The shape of the lens aperture with a single vane and fixed mask: (a) the mask alone, (b) at $f/2.8$, (c) at $f/4$, and (d) at $f/11$.

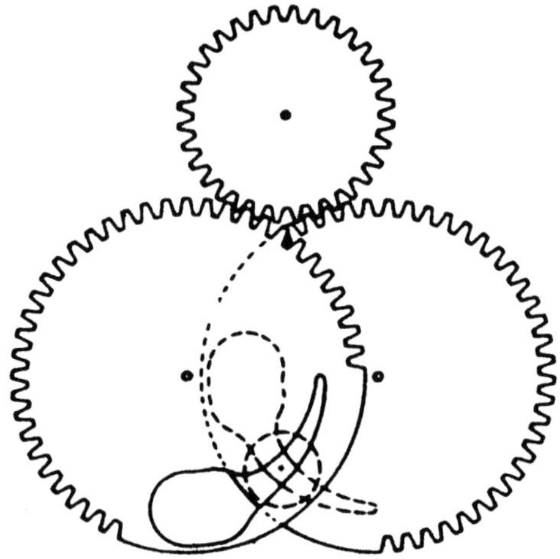

Figure 6.8. Double-vane automatic aperture control.

photocells controlling the exposure are often mounted on these "ears," thus avoiding further loss of light. Even this small loss can be avoided by using two disks carrying spiral slots perpendicular to each other, which are coupled together by very delicate gearing so that the aperture is always a close approximation to a square (Fig. 6.8). Both of these arrangements have been used in small movie cameras.

The Variation of Illumination over the Field of a Lens

Vignetting

In most photographic lenses, the finite length of the barrel between the entrance and exit apertures causes the oblique rays to be gradually cut off as the obliquity is increased, until eventually the light is extinguished altogether. This effect is clearly illustrated in Fig. 6.9, and it exists in all lenses that are required to cover a field of more than a few degrees in extent. A normal camera lens covering a semifield of 26°, for example, will have an effective aperture at the corners of the film that is less than half the aperture at the center of the picture, on an area basis. Provided this vignetting of oblique pencils is not much more than 50%, however, it is not likely to be too serious. It has the effect of somewhat darkening the corners of a color transparency and lightening the corners of a negative. Reduced density in the corners of a negative can be offset by suitable

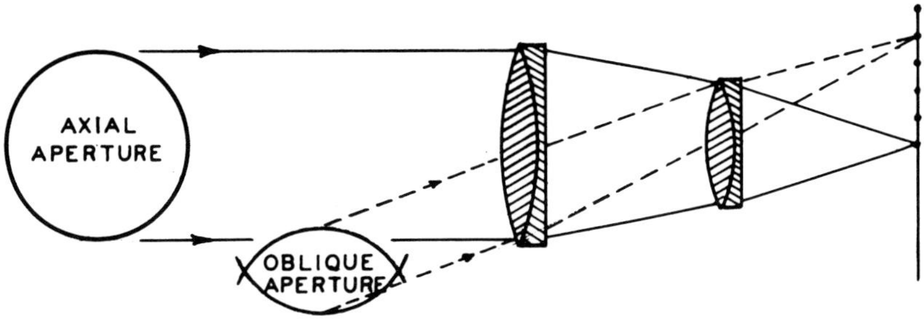

Figure 6.9. The vignetting of an oblique light beam in a lens.

dodging during printing, and it is automatically compensated to some extent by the vignetting in the enlarger lens.

The Cos⁴ Law

Even if the front and rear apertures of a lens are made so large that they do not in any way limit the oblique rays passing through the lens, the illumination on the film will still diminish in the outer parts of the picture. There are three distinct cosine factors that tend to reduce the illumination in the image at the corners of the picture, as may be seen in Fig. 6.10. First, at an obliquity ϕ the circular aperture of the lens appears to be an ellipse, the area being reduced by factor $\cos\phi$. Second, the light in the camera falls upon the film at an angle ϕ, which again causes a reduction in illumination by $\cos\phi$. Third, the distance from the lens to the film at the obliquity ϕ is greater than the axial distance, by a factor $\sec\phi$, leading to two more $\cos\phi$

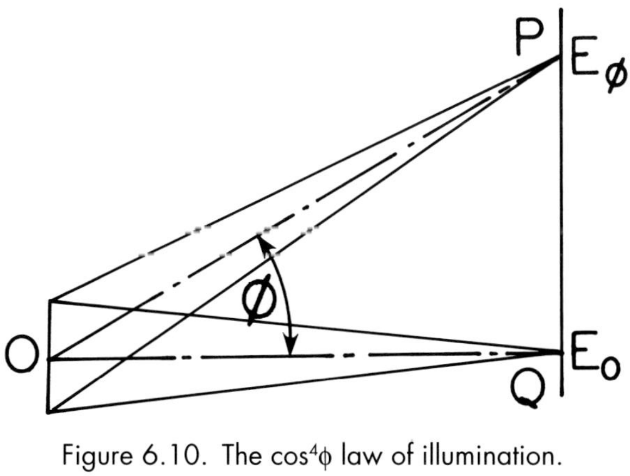

Figure 6.10. The $\cos^4\phi$ law of illumination.

factors by the operation of the ordinary inverse square law. These four cosine factors all tend to reduce the corner illumination, with the result that the ratio of the image illumination at angle ϕ to the central illumination is given by

$$E\phi = E_0 \cos^4 \phi \ . \qquad (6.13)$$

This law operates in all lenses regardless of whether or not there is any vignetting of oblique light by the lens barrel. Consequently, when both effects are present together, the image illumination may diminish very drastically in the corners of the picture area. A partial cure may be obtained by closing down the lens iris to eliminate the vignetting, but this will in no way affect the $\cos^4\phi$ factor. Some typical $\cos^4$ values are given in the following table:

ϕ	0	5°	10°	15°	20°	25°	30°	35°	40°	45°	50°
$\cos^4\phi$	1.00	0.98	0.94	0.87	0.78	0.67	0.56	0.45	0.34	0.25	0.17

There are a few unusual wide-angle lenses in which the drop in illumination from center to edge of the field differs markedly from the value given by the product of the vignetting fraction and $\cos^4\phi$. This is because the front component of the lens distorts the diaphragm at high obliquities, thus admitting either more or less light than would be admitted by a circular undistorted diaphragm. A typical lens that gives less light than the $\cos^4$ law indicates is the Metrogon (Topogon). On the other hand, any objective having a strong negative front component, such as the Aviogon (Biogon) or any reversed-telephoto lens, will give significantly more oblique illumination than would be expected on the basis of the $\cos^4\phi$ formula.

It should also be mentioned that image distortion can also affect the distribution of illumination across the picture area. If a lens has considerable barrel distortion, as is the case in a fish-eye lens, the crowding of the oblique images increases the illumination there, making it more nearly equal across the film. On the other hand, if a lens has pincushion distortion, the spread of the available light over a greater area at the edge of the image leads to a reduction of the illumination there.

Thus, there are four causes of unequal illumination across the image in a camera: (a) vignetting, (b) the $\cos^4$ law, (c) pupil distortion, and (d) image distortion. As a result of all these defects, one cannot readily deduce the actual variation of illumination in any particular case without making actual trials or extensive ray-trace calculations.

Light Reflection in Lenses

It was shown in Chapter 3, page 59, that polished glass surfaces reflect light to an extent depending on the refractive index of the glass itself. This simple fact is at the bottom of a whole range of objectionable phenomena known as ghost images, which occasionally spoil an otherwise good photograph.

The reason why surface reflection in lenses is so serious is because a lens may have anywhere from four to ten or more glass-air surfaces. If each surface of an eight-surface lens reflects 5% of the incident light, a simple calculation reveals that the transmittance t of the lens should be equal to $(0.95)^8$, or 66%.

If this reduction in transmitted light flux were the entire story, we should not be too much concerned because we could use a longer exposure time to compensate for the low lens transmittance. Unfortunately, the light reflected from the inner surfaces of the lens must find its way out again through the surfaces previously encountered, and in so doing a few percent of the light will be reflected again, this time back into the camera where it falls on the film as unwanted illumination (Fig. 6.11). For a two-surface lens, the intensity of this light is only 5% of 5% or one-quarter of 1%, which is quite insignificant, but in a complicated lens it may become serious. For instance, in an eight-surface lens there is the possibility that 7 + 6 + 5 + 4 + 3 + 2 + 1 = 28 beams of doubly reflected light will find their way to the film, and even though each beam may have an intensity of only 0.25% of the original light, yet 28 times 0.25% amounts to no less than 7%. The exact value is not quite as high as this because of the reduction in the intensity of each beam as it is transmitted through each lens surface. A more precise calculation shows that if r is the reflectance of each surface and N is the number of surfaces, the total transmittance of the lens including all possible multiple reflections, but ignoring any loss of light by direct absorption in the glass, will be given by*

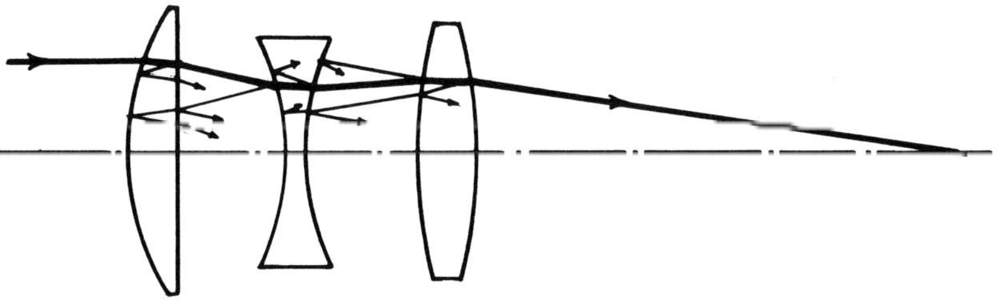

Figure 6.11. Some of the doubly reflected beams of light in a lens.

*R. A. Hull, "Transmission of light through a pile of parallel plates," Proc. Phys. Soc. **48**, 574–575 (1936).

$$t = \frac{1 - r}{1 + (N - 1)r} \quad . \tag{6.14}$$

This formula may be compared with that for the amount of useful light transmitted by the lens, namely,

$$t = (1 - r)^N \quad . \tag{6.15}$$

A series of values of both formulae are given in the following table:

For refractive index 1.57, r = 0.05				
Number of glass-air surfaces	**Useful light transmitted**	**Total light transmitted**	**Unwanted light**	**Ratio of unwanted to useful light**
2	90.2%	90.5%	0.2%	0.2%
4	81.4%	82.6%	1.2%	1.4%
6	73.5%	76.0%	2.5%	3.4%
8	66.3%	70.4%	4.0%	6.1%
10	59.9%	65.5%	5.6%	9.4%

The important quantity for the practical photographer is the glare index given in the last column, because this measures the loss of contrast that must be expected as a result of internal reflections within the lens.

Ghost Images

It is a mistake to suppose that the unwanted multiply reflected light referred to in the above section will be spread uniformly over the entire film area. Since each surface acts as a curved mirror, each reflected beam forms somewhere in space a reasonably well-defined image of every object situated in front of the lens, and because most lens surfaces are strongly curved, there is a tendency for all the doubly reflected images to be formed either inside of, or very close to, the lens itself. They may be readily observed by removing the back of a camera and looking into the lens from behind while directing the camera toward a strong lamp (Fig. 6.12). However, in some unfortunate cases, one of these reflected images may happen to be projected far out behind the lens, and it may fall so close to the film as to form upon it a sharp or only slightly blurred image of a distant bright object (Fig. 6.13). This is called a *ghost image*, and it is of course an

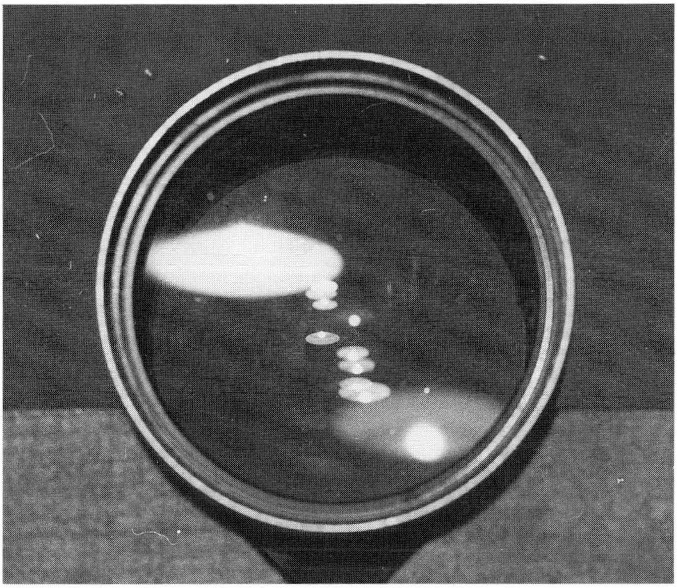

Figure 6.12. The reflected images of a lamp, seen from the front of a photographic lens. (This lens shows eleven reflections, eight from the glass-air surfaces and three from the cemented surfaces. Because the lens is coated, the glass-air surface reflections are no brighter than those from the cemented surfaces.)

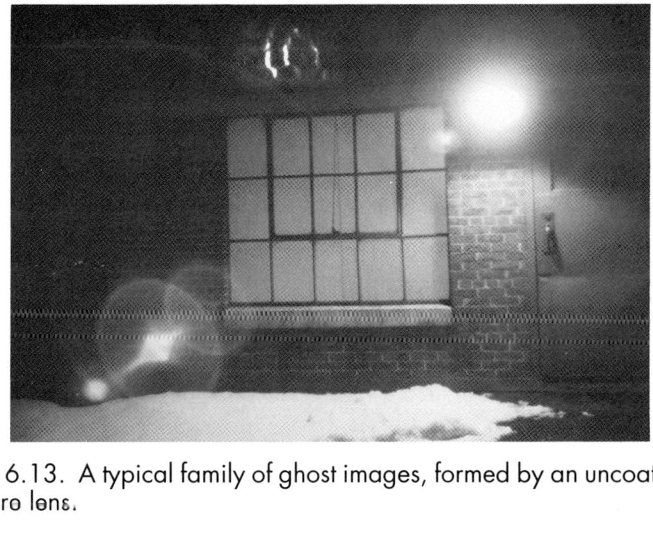

Figure 6.13. A typical family of ghost images, formed by an uncoated high-aperture lens.

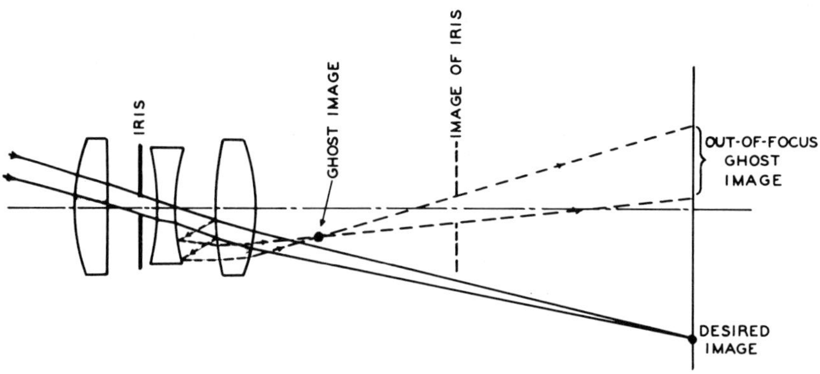

Figure 6.14. The formation of ghost images by light reflected from the internal surfaces in a lens.

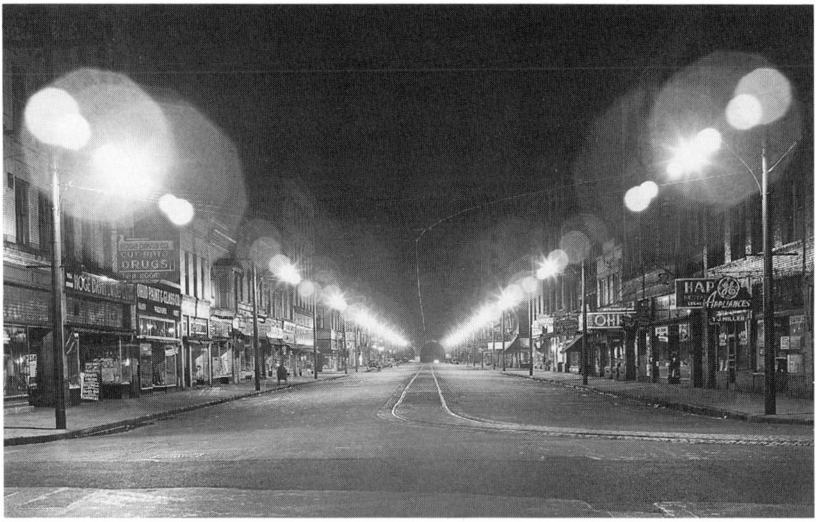

Figure 6.15. A typical series of out-of-focus ghost images, appearing as projections on the iris diaphragm.

extremely annoying phenomenon. A lens hood (see page 166) helps by preventing direct sunlight from falling on the lens, but some ghosts are formed by bright lights actually within the lens field where a sunshade or lens hood cannot be of any help.

If a ghost image of a small bright lamp or other source of light is out of

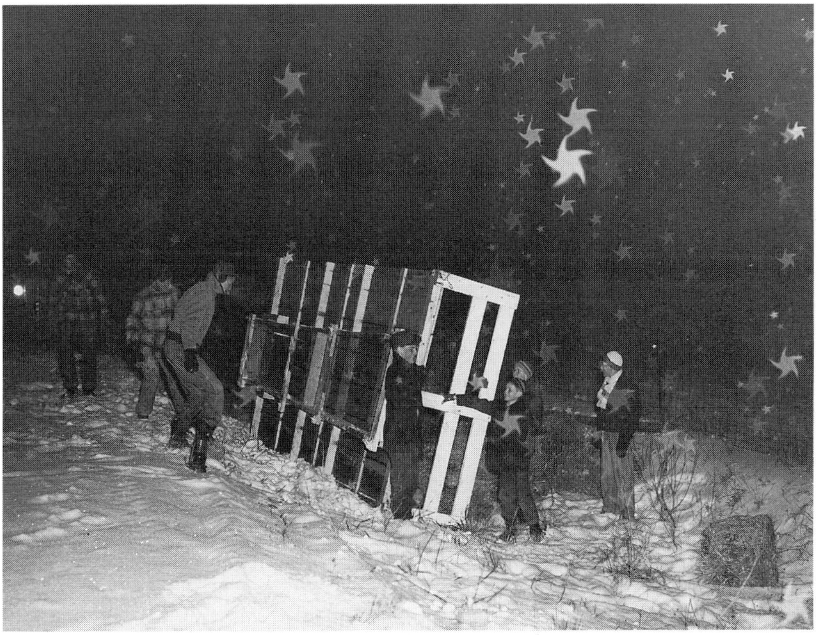

Figure 6.16. Out-of-focus ghost images of snowflakes projected onto the film through a poorly synchronized, partly open shutter.

focus, and the lens is stopped well down, the converging reflected light beam that forms the ghost will have a cross-section similar to the actual shape of the iris diaphragm (Fig. 6.14), and such an out-of-focus ghost image will look like a projection of the iris itself on the film. This type of ghost is common with some forms of lens, a serious example being that illustrated in Fig. 6.15. The same phenomenon has accounted for the numerous images of the shutter opening scattered over the picture in Fig. 6.16 taken during a snowstorm. The flash was poorly synchronized, causing it to operate before the shutter was fully open, and each snowflake close to the camera was projected as an out-of-focus ghost on the film.

Flare Spot

As shown in Fig. 6.14, any doubly reflected light beam from a lens will cross the lens axis somewhere, and at that point it will form a well-defined image of the iris diaphragm itself. If this happens to fall close to the film plane, it will cause a more or less sharply defined image of the iris, superposed on the

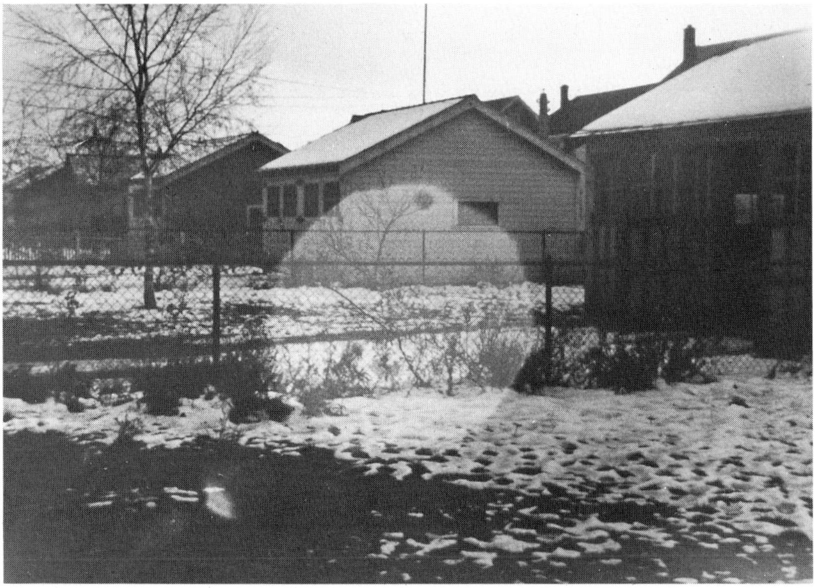

Figure 6.17. An in-focus diaphragm image on the film (enlarged from a 16mm frame).

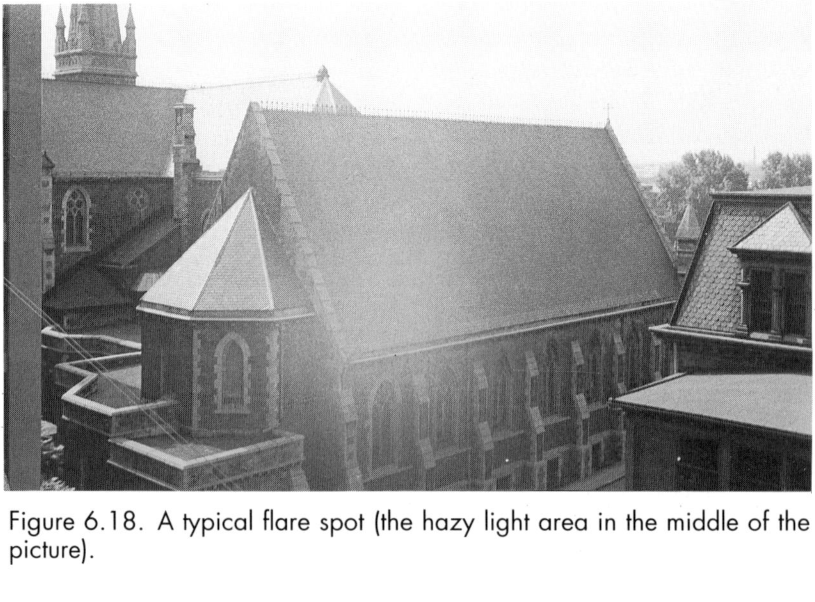

Figure 6.18. A typical flare spot (the hazy light area in the middle of the picture).

middle of the picture (Fig. 6.17). This may be easily distinguished from the out-of-focus ghost image referred to in the last section because it is precisely central, while any kind of ghost image moves about as the source is moved in the field.

If several of these multiply reflected images of the iris diaphragm fall close to the film but not in focus upon it, their intensities may nevertheless add up to a considerable amount of light, causing a *flare spot* in the middle of the picture (Fig. 6.18).

A flare spot, or an in-focus central diaphragm image, is visible only at the smallest diaphragm openings, because it is usually highly magnified, and at the larger diaphragm openings the boundary of the image falls entirely outside the picture area. Moreover, the brightness of the diaphragm image is independent of the size of the iris, while the brightness of the background increases as the square of the iris opening; hence, as the diaphragm is opened, the background brightens rapidly and soon drowns the faint iris image entirely. The color of a flare spot is, of course, the average of the whole scene.

A flare spot is particularly unfortunate when stray light reflected from shiny regions in the lens mount is also present, because the two sources add up and may lead to a serious situation. Indeed, many cameras have been entirely cured of their flare-spot troubles by an adequate blackening of the interior of the lens barrel.

Lens Coating

For many years lens manufacturers have sought some means for reducing the surface reflectivity of glass, because if such could be found all the above reflection troubles could be eliminated at one stroke. The possibility of this was recognized by H. Dennis Taylor in 1896 when he noticed that some old lenses having dark tarnished surfaces transmitted more light than a new lens of the same kind, the reduction of surface reflectivity due to the tarnish actually causing an increase in the lens transmittance. Subsequent attempts to produce this kind of tarnish artificially were irregular and generally unsatisfactory, and the procedure was never attempted on a commercial scale. In 1936, it was suggested* that surface reflection could be reduced by depositing a thin layer of some low-index material on the

* J. Strong, "On a method of decreasing the reflection from non-metallic substances," J. Opt. Soc. Am. **26,** 73–74 (1936). The same procedure was patented in Germany by A. Smakula in 1935 (D. R. P. 685,767).

surfaces of a lens. The material first proposed was calcium fluoride, and the deposition was done in a high vacuum by direct sublimation of the heated crystalline powder. As the thin layer is gradually deposited in the vacuum chamber, it passes through the same range of colors that are seen on the surface of iron when it is strongly heated, namely, light straw, darker straw, brown, magenta, blue, and finally deep purple. The correct point to stop the deposition for normal photographic purposes is somewhere around the brown stage.

The action of the thin film in reducing surface reflection is an interference phenomenon similar to that which occurs in a soap bubble. Part of the incident light is reflected from the upper surface of the film (Fig. 6.19), and the rest passes through the film to the film-glass interface. Here a second part of the light is reflected and the remainder transmitted. If the film thickness is such as to cause a half-wave lag between these two reflected beams, and if the two beams are equally bright, they will interfere with each other and eliminate the reflected light entirely. However, there will usually be some faint reflected light from an antireflection coating

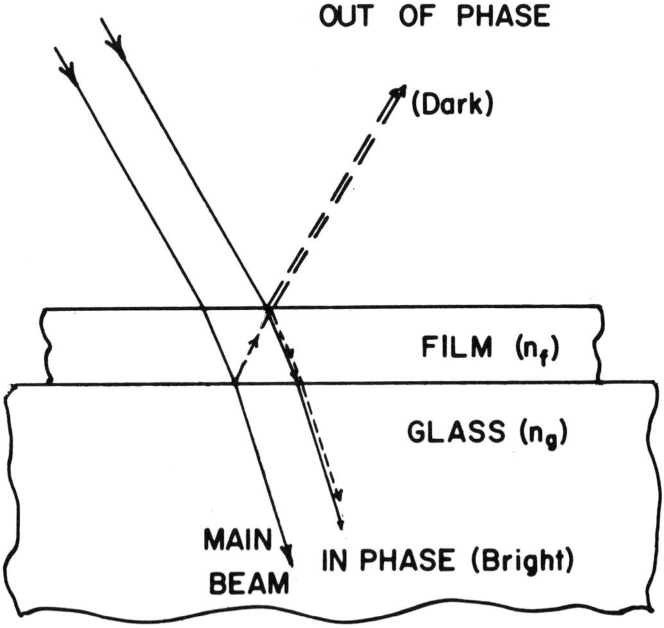

Figure 6.19. Illustrating the effect of coating the surfaces of a lens.

unless the refractive index of the coating material is equal to the square root of the glass index, in which case the intensities of the two interfering beams are equal.

Commercial low-reflection coatings of this kind were first offered for sale in December 1938, and since that time great strides have been made in improving the process. Calcium fluoride is very effective in reducing reflection, but it yields a soft layer that is easily rubbed off. Consequently, it was soon replaced by other fluorides, which, together with improvements in the vacuum technique, have made possible coatings that are now as hard and durable as the glass itself. The photographer is cautioned, however, against the use of silicone-treated cleaning cloths, as the silicone material changes the characteristics of a coated lens.

The spectral distribution of reflected light from a single-layer antireflection coating generally drops to a minimum somewhere in the visible spectrum and rises at wavelengths on each side of the minimum. Greatly improved reduction of reflection can be achieved by the use of a multilayer coating, in which a first layer of a high-index material is followed by a layer of low-index material, and, if everything is done properly, a decidedly improved reduction of reflection throughout the entire visual spectrum can be obtained. Many of today's better lenses are multilayer coated for this reason, and ghost images are virtually eliminated even when the sun or other bright light source happens to fall within the picture area.

Advantages of Lens Coating

As has been explained, coating the surfaces of a lens with a reflection-reducing layer has two principal effects: (a) it greatly reduces the brightness of ghosts and flare spots, so that they are rarely seen under normal conditions, and (b) it increases the light transmitted by the lens in the highlight region and reduces it in the shadow region.

The value of the first of these effects goes without question, although we should realize that a strong light inside or just outside the field of view may cause a ghost even with coated lenses, because the surface reflection is only reduced and not entirely eliminated. However, the second effect calls for further consideration.

If the "speed" of a lens-film combination is judged by the density in the highlights, as is common practice in the professional motion-picture industry, then coating a lens can increase its effective speed considerably. An eight-surface lens, for example, which transmits 65% when uncoated, may easily transmit 90% when coated, representing a gain of 38% in speed, which is equivalent to nearly half a stop. On the other hand, if the speed of a lens-film combination is judged by the density in the shadows, which is standard practice in ordinary black-and-white photography, then coating a lens may actually reduce its speed by removing from the shadows the unwanted multiply reflected light from the lens. The net result of increasing the light in the highlights and reducing the light in the shadows is to increase notably the contrast of the image. Since the unwanted reflected light is, in general, of a different color from that of the image itself, coating may also lead to a significant improvement in the color rendering.

The advantage of coating shows up most noticeably when photographing a person or other low-contrast subject in front of a bright surface such as an open window. The unwanted reflections from the bright window tend to lower the contrast of the image of the person so seriously that, in bad cases, only a silhouette of the person is obtained. The major benefit of lens coating is seen in all such conditions as this; consequently, when equipped with a coated lens, the photographer does not have to stop and consider whether or not a background is bright enough to dull the subject contrast, but can forget this possibility entirely and go ahead with the picture.

Other Sources of Unwanted Light

The chief secondary sources of unwanted, or stray, light in an image are shiny areas inside the lens barrel and dirt or scum on the lens surfaces.

Barrel reflections are almost always eliminated in modern lenses by the judicious addition of knife-edged baffles in the barrel (Fig. 6.20). Although these baffles do not prevent light from being reflected from the inside of the barrel, they catch the reflected light before it can reach the film. The inside of the barrel of a lens equipped with well-placed baffles looks absolutely black when the back of the lens is viewed from a point inside the area of the film aperture. However, it is fairly common to see some shiny areas in the barrel of a cine lens, for example, when viewed from a point well outside the film area, but this, of course, is quite unimportant. To be effective, a baffle must be knife-edged and blackened so that no light is reflected or scattered from the edge of the baffle itself. Merely blackening, corrugating, or roughening the inside of a lens mount is usually not sufficient to prevent barrel reflections.

Even the ground edges of the lens elements themselves may be a source

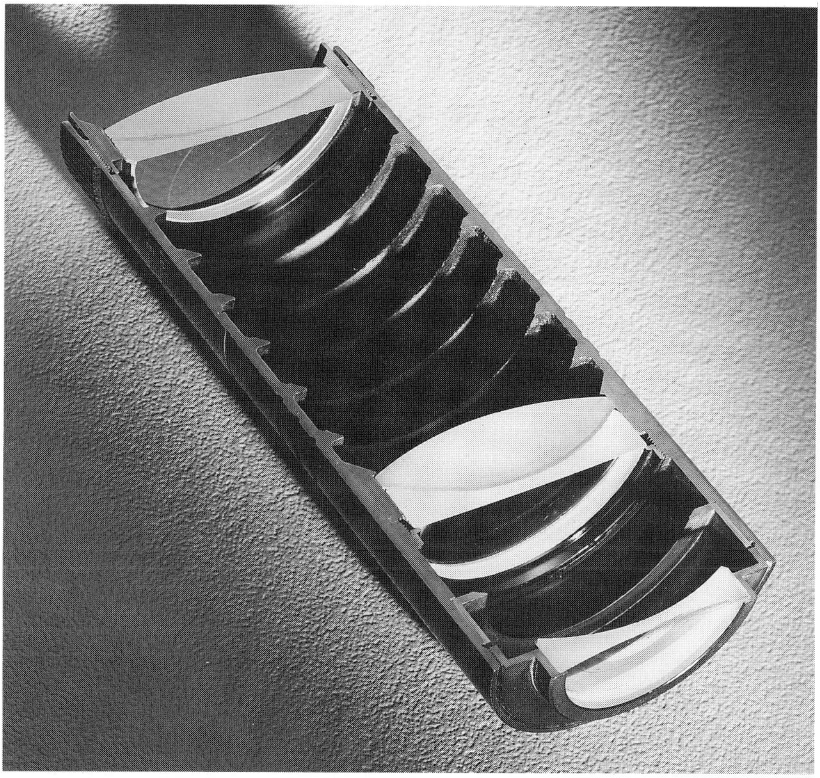

Figure 6.20. The use of baffles to eliminate barrel reflections in a lens.

of stray light. If the edge of a lens can be seen from a point in the film aperture, then it should be blackened before assembly to eliminate reflections.

Dirt and scum on the surfaces of a lens are obvious causes of poor contrast in the image. The extent of the reduction in contrast caused by a small amount of dirt is surprising, and often not realized by the user. A strong thumbprint on a lens is sufficient to lower the contrast appreciably. The remedy is to keep a lens clean, for repeated cleaning will only result in fine scratches on the lens surfaces.

The inside surfaces of a camera bellows, and more seriously the inside of the small plastic camera, can also act as a prolific source of unwanted light on the film. This situation can be helped considerably by the use of an adequate sunshade over the lens.

Projection Screens

Reflection Screens

The most commonly used reflection screen is a simple white-painted cloth, sheet, or wall. As this closely approximates to a perfect diffuser, its brightness will be equal to kE/π, where k is almost unity, and it will appear to have this brightness when viewed from any direction. If enough light is available in the projection system, this type of screen is unsurpassed.

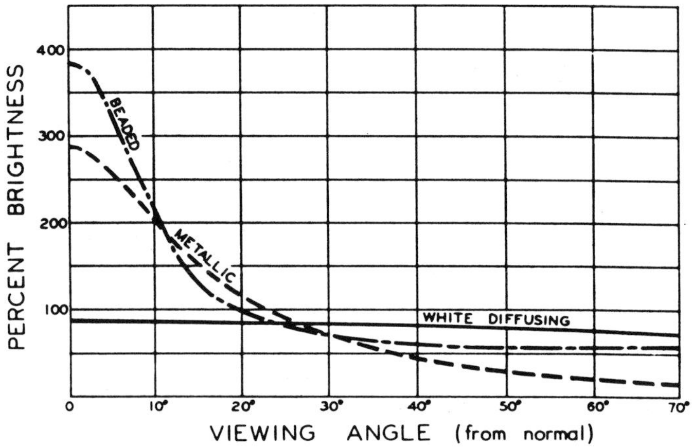

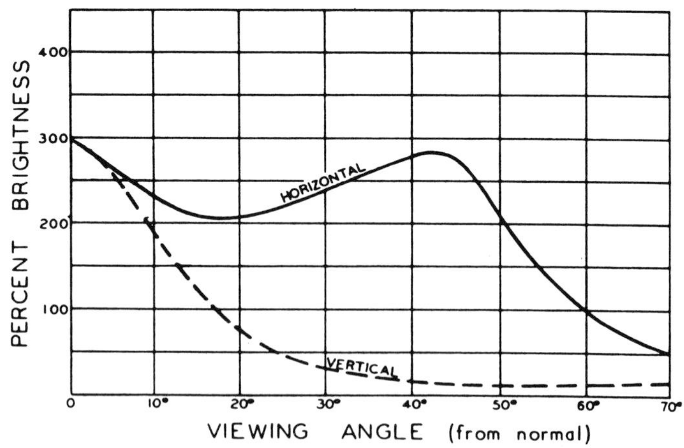

Figure 6.21. The angular variation in brightness of different types of projection screens. The lower graph refers to a lenticular screen with vertical grooves.

However, it often happens that there is insufficient light available to produce a satisfactory image on a plain white diffusing screen, especially if there is much ambient illumination also falling on the screen. The remedy is to use a partially specular screen such as sand-blasted aluminum or one coated with metallic paint, a "beaded" screen covered with tiny glass spheres on a metallized backing, or a lenticular screen having a highly specular plastic coating that is ribbed vertically so as to cause a controlled horizontal spread with very little vertical spread. In Fig. 6.21 the diffusing properties of several screens are shown in graphical form. The older types are indicated in the upper chart, whence it is clear that the gain in brightness is entirely lost if the observer is more than 20° away from the incident beam. The lower chart represents the measured horizontal and vertical spread of a modern lenticular screen.

Transmission Screens

The most perfectly diffuse material for a transmission screen is opal glass, but its transmittance is so low that the image is likely to be too dim for comfortable observation. Some successful diffusing screens have been made by depositing chemical substances on glass, such as those provided with Recordak Film Readers. Plain ground glass is more common, but its diffusing properties are often insufficient to cause the more oblique incident light to be scattered toward the observer's eye (Fig. 6.22). The

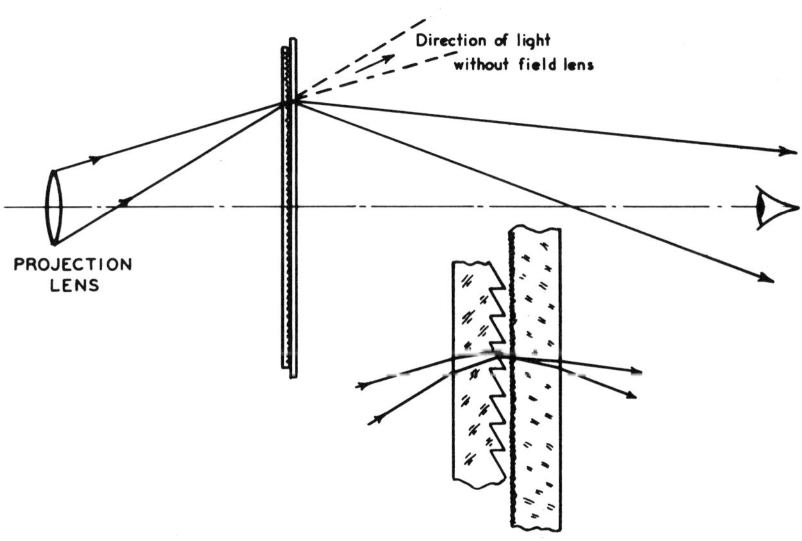

Figure 6.22. A flat plastic Fresnel lens behind a ground-glass screen.

result is the formation of a bright "hot spot" surrounding the point on the screen where it is pierced by a line joining the projection lens to the observer's eye, and as the observer's head moves laterally, the hot spot follows. Outside the bright hot spot, the screen appears relatively dim.*

In the limit, of course, if *all* diffusion were removed and the screen were replaced by a clear glass sheet, the hot spot would become nothing but the projection lens aperture, and its brightness with no slide in the projector would be equal to the brightness of the projection lamp itself. Diffusion, then, increases the size of the hot spot at the expense of image brightness.

As was mentioned in Chapter 2, it is possible to significantly increase the size of the hot spot by the addition of a field lens to the ordinary ground-glass screen. A simple plano-convex lens can be used, but a flat plastic Fresnel lens will be found more convenient (Fig. 6.22). Indeed, if the eye is correctly located within the image of the projection lens aperture formed by the field lens, the diffusion may be dispensed with altogether to give a fully specular system. This results in an astonishing increase in the picture brightness, and the entire field will appear as bright as the lamp filament itself! However, the latitude in eye position is then likely to be too small for convenient observation, and some diffusion must be employed to make a satisfactory system.

It has long been realized that a neutral density filter placed in front of a translucent diffusing screen greatly increases the image contrast when there is considerable ambient illumination. This is because the useful light has to pass through the neutral density once, whereas the room light must pass through it twice before reaching the eye. For this reason television picture tubes are generally made with a slightly absorbing end face, the gain in contrast more than offsetting the loss in image brightness caused by the absorption in the glass.

Photographic Density

The term density as applied to photographic emulsions and other absorbing media is defined as the logarithm of the reciprocal of the transmittance, which is the ratio of the emerging flux to the incident flux. Hence, density is defined by

$$D = \log_{10} \frac{1}{t} . \tag{6.16}$$

*P. Vlahos, "Selection and specification of rear projection screens," J.S.M.P.T.E. **70,** 89–95 (1961).

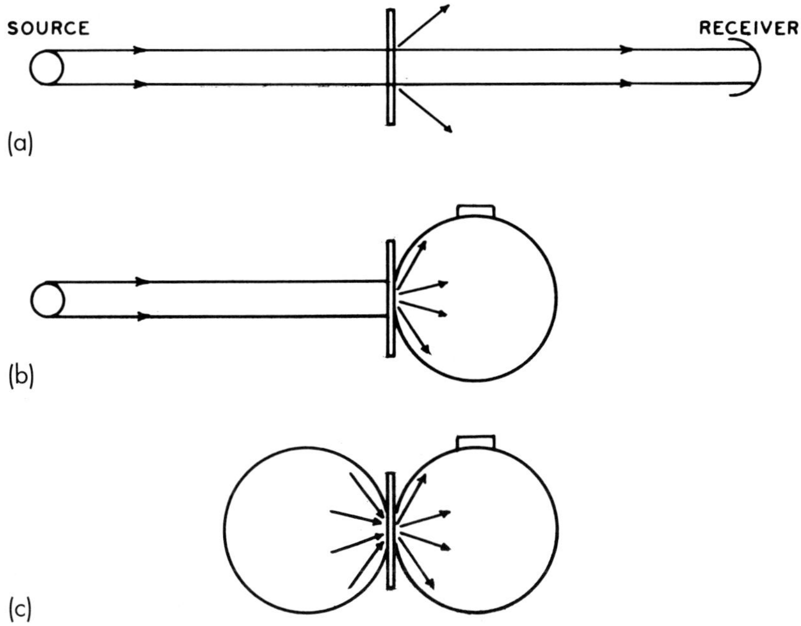

Figure 6.23. The three types of density measurement: (a) specular, (b) diffuse, and (c) double diffuse.

The following table gives some typical values of density and the corresponding transmittance values:

D	*t*
0	100%
0.1	79
0.2	63
0.5	32
1.0	10
1.5	3.2
2.0	1.0
2.5	0.3
3.0	0.1

When we attempt to measure the density of an emulsion, some further qualifications are necessary because the silver grains in the emulsion partly absorb and partly scatter light, and we shall obtain different values for the density depending on whether we include or exclude this scattered light. The difference between these two values, called diffuse and specular density, respectively, may be illustrated graphically by Fig. 6.23. If the source is distant so as to supply the test film with roughly parallel light, and if the measuring apparatus is also distant as shown in Fig. 6.23(a), we shall exclude all the scattered light and obtain a measure of the specular density of the specimen.

On the other hand, by using an integrating device close to the film that is capable of collecting up all the light radiated by the film into a complete hemisphere, as in Fig. 6.23(b), we shall measure the diffuse density, which is obviously lower than the specular density since more light is apparently transmitted by the sample. It should be noted that the situation in Fig. 6.23(b) may be reversed; that is, we shall obtain a valid measurement of the diffuse density of a specimen if we send light into the sample from all directions and measure the emitted light by means of a distant receiver.

We may go still further and use both hemispherical incidence and hemisphere collection, as indicated in Fig. 6.23(c). This gives us a measure of double-diffuse density, which actually turns out to be somewhat higher than the ordinary diffuse density, mainly because of the longer path that the obliquely incident light must take through the emulsion.

It should be emphasized that specular and diffuse densities are both limiting cases in which the collection cone-angles are zero and 180°, respectively, for normal incident illumination. If the collector takes in light at some intermediate angle, we should expect to find a value of density intermediate between the true specular and true diffuse density. Actually, it requires a total collecting angle of only about 120° to reach a density substantially equal to the true diffuse density, because in a developed photographic material most of the scattered light is diffused into quite a small angle. The effective density of a transparency in any practical projector or condenser-type enlarger, therefore, lies between its specular and diffuse density, in spite of the fact that a so-called specular beam of light is used. The term density, without qualification, is usually taken to mean the diffuse density. True specular density, in which no scattered light is included is almost never encountered in practice.

The double-diffuse density of an emulsion is significant only in contact printing from a broad diffuse source such as the open sky that sends light to the negative from all directions.

Callier's Q-Factor

The ratio of the specular to the diffuse density of an emulsion is known as *Callier's Q-factor*. For an ordinary negative material it has a value approaching 1.6. The Q-factor was at first thought to be a fixed property of a given emulsion, but this was later shown to be untrue,* and actually the value of Q varies considerably with both density and gamma.

Reflection Density

In measuring the blackness of a photographic print, the term reflection density is often encountered. This is a somewhat specialized concept, since the glossy surface of many types of photographic paper tends to reflect light like a mirror, and this light must be excluded from consideration.

In measuring reflection density, therefore, it is customary to illuminate the sample with light at about 45° to the surface and measure the diffusely reflected light in a direction perpendicular to the surface. The standard of 100% reflectance (i.e., zero reflection density) is taken to be the unexposed fixed-out surface of the paper base. Then the logarithm of the ratio of the light from the paper base to the light from an equal area of the specimen is defined as the reflection density of the specimen.

---

*The whole subject of density and its measurement has been covered by C. E. K. Mees in *The Theory of the Photographic Process*, 4th ed; T. H. James, ed., Macmillan, New York (1977).

Chapter 7

Types of Photographic Objectives

Many types of construction have been adopted in photographic lenses since the invention of photography in 1839, but very few of these have survived to the present day. The simpler types generally work at low relative apertures, but the supposed rule that a high aperture and a wide angular field are incompatible no longer holds. As was mentioned in Chapter 1, the "normal" lens on a still camera covers a field of about $\pm 24°$, and today such lenses can be made with an aperture of *f*/1.4, while a few special lenses attain an aperture as high as *f*/1 when the diameter of the lens opening is equal to the focal length. A few special lenses have been made with an aperture greater than this, but they mostly cover only a very narrow field.

At one time, not too long ago, photographers were very much aware of the type and kind of lens they were using on their cameras, and it was a matter of pride to own a Rapid Rectilinear or some type of anastigmat. Gradually, however, knowledge of lens structure has become less important, and today most photographers have absolutely no idea of or interest in the structure of their lenses. So long as the lens has the desired focal length and aperture, is priced within their means, and gives good definition with adequate depth of field, most photographers are satisfied. This is not a unique phenomenon, as most people have no knowledge of the working of their refrigerators, calculators, typewriters, or of many other appliances in everyday use. However, there are some who are interested in lenses, and it is hoped that this chapter will give them some useful information.

The camera manufacturer, of course, is vitally concerned with lens structure, and endeavors to obtain the best possible image quality at the lowest cost. Modern methods of lens design and manufacture are of the greatest importance to the entire camera industry.

Lens design today has been enormously helped by the coming of the high-speed electronic computer. Indeed, many of the modern reversed telephoto and zoom lenses would have been impossible to design without it. When computers first appeared in the early 1950s, it took several seconds to trace a single ray through a spherical refracting surface. Today, using the largest computers, a speed of about 50,000 ray-surfaces per second is possible, representing an increase in speed of an order of magnitude every 4 or 5 years, and the end is not in sight. At these enormous speeds it is possible to program the computer to make all possible changes in a design and select those changes that effect an improvement in the lens, so that after only a few seconds of operation any given lens design can be improved to its optimum limit. Thanks to these computer optimization programs, lenses today are vastly better than ever before, in both aperture and performance.

Obviously, lenses must now be manufactured much more accurately than before if this improved design capability is to be realized. The sections of two actual lenses shown in Fig. 7.1 indicate how complicated lenses have now become, and particularly how elaborate is the lens mounting. Mount details include, of course, the iris diaphragm and the focusing thread, but in addition there must be coupling linkages by which the iris can be operated by the exposure control mechanism in the camera, and in some cases the focusing adjustment must be coupled to the camera rangefinder

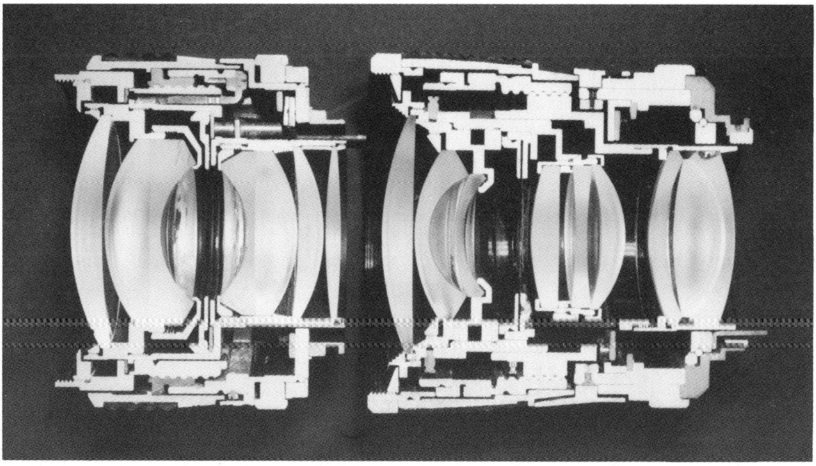

Figure 7.1. Sections of two typical modern photographic objectives, showing the complexity of the mounting

or automatic focusing mechanism. A zoom lens must embody all these features in addition to the complex mechanism required to permit a continuously variable focal length.

For those who may have some curiosity as to how lenses are constructed, the remainder of this chapter contains a description of the principal types of present-day camera lenses and why that particular construction is employed.

Most of today's photographic objectives fall into seven easily recognizable classes:

(a) low-aperture singlets or periscopics,
(b) medium-aperture triplets or Tessars,
(c) high-aperture double Gauss or Sonnar types,
(d) telephoto lenses,
(e) reversed telephotos,
(f) zoom lenses,
(g) catadioptric and mirror systems.

Besides these, there are numerous special types of lens, for aerial photography and photogrammetry, projectors, enlargers, copy cameras, optical filtering and image processing, and so on. Most of these will not be discussed here.

Low-Aperture Singlets

A typical single lens is the common landscape lens used on low-priced box cameras (Fig. 7.2). This lens was first suggested by W. H. Wollaston in 1812 for use on camera obscuras, some twenty years before the invention of photography. The field of such a lens can be flattened by making it of a suitable meniscus shape, and the coma corrected by placing the stop at the right distance from the lens. At apertures of *f*/14 or less, this lens permits satisfactory contact prints or low degrees of enlargement over a semi-field of about 25°. When stopped down, however, excellent negatives can be obtained. A more recent development is to turn the lens around so that it

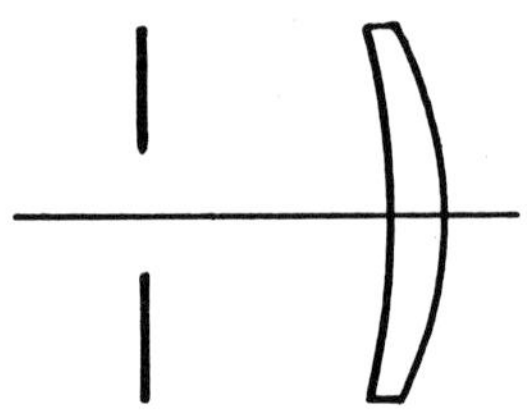

Figure 7.2. The Wollaston landscape lens.

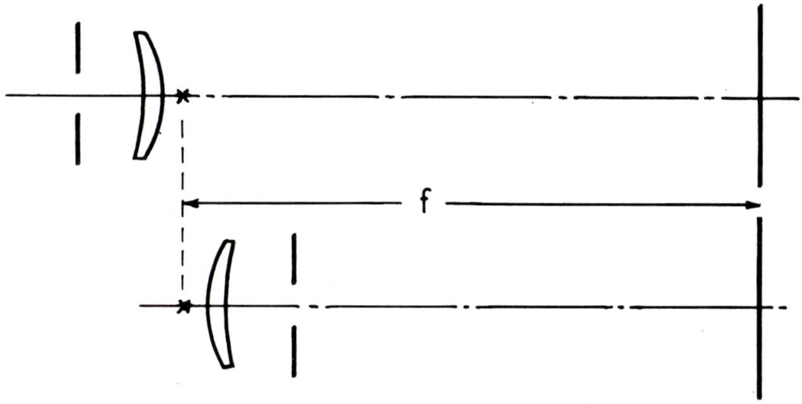

Figure 7.3. Mounting the stop behind the lens shortens the camera.

comes in front of the stop instead of behind it. This arrangement leads to a considerable shortening of the camera (Fig. 7.3) and provides an excellent cover to keep dirt, sand, etc., out of the shutter mechanism. As the field is likely to be somewhat curved, it is generally advisable to give the film a cylindrical form in the camera to offset in part the field curvature of the lens.

Some attempts have been made to achromatize the landscape lens, and the early French Daguerreotype cameras were equipped with such a lens as suggested by C. Chevalier [Fig. (7.4)]. However, the practical advantages of this achromatization are not sufficient to justify the increased cost, and the type has now been almost entirely abandoned. Nevertheless, photographers are often surprised at the great improvement in picture sharpness that can be gained by the use of a yellow filter with a box camera, due to the reduction in chromatic aberration and lateral color.

Sometimes a pair of identical simple menisci are mounted symmetrically about a central stop; such an arrangement is known as a *periscopic lens*. Such lenses benefit from all the advantages conferred by symmetry (see

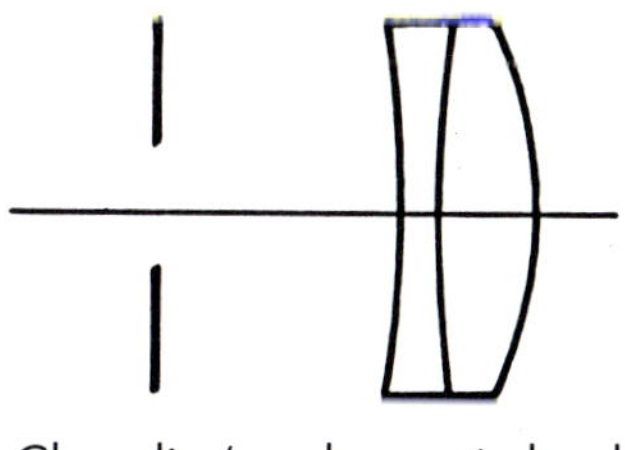

Figure 7.4. Chevalier's achromatic landscape lens.

page 48), but they still suffer from spherical and chromatic aberration, requiring a low aperture such as *f*/15, and they are about twice as expensive as a single meniscus. For this reason they are seldom used today in low-cost cameras. However, lenses of the periscopic type are often employed on vuegraphs for lecture presentations, the image of the small lamp serving to stop the lens down so that its spherical aberration is insignificant.

Medium-Aperture Lenses

Medium-aperture lenses work at an aperture of *f*/6 to *f*/8 and usually contain three elements (Fig. 7.5). This so-called Cooke Triplet lens was originated in 1893 by H. Dennis Taylor, chief designer of T. Cooke and Sons of York, England. As Cooke did not wish to make photographic lenses, the design was submitted to the small optical firm of Taylor, Taylor and Hobson in Leicester, but out of courtesy to Dennis Taylor's employer, the new lens was called the *Cooke Triplet*. At first it was made as a high-quality portrait lens, but gradually it became adapted to more moderately priced cameras and today it is the standard type of construction for all medium-aperture (*f*/6 or *f*/8) photographic objectives.

Almost any selection of glass can be used, provided there is sufficient dispersion difference between the crown and flint elements, and the higher the refractive indices the greater the possible aperture. Today, many Cooke Triplet lenses are made of plastic materials, and some recent designs have one or more aspheric surfaces, permitting an aperture of *f*/2 or even higher. However, the high temperature coefficient of refractive index of plastic materials may limit the usefulness of these high-aperture lenses unless some means is incorporated to maintain the image in focus over a wide range of ambient temperatures. One such procedure embodies a differential thermal arrangement to increase the front airspace at elevated temperatures when the focus would otherwise fall beyond the film plane; an alternative procedure is to use plastics for only the front two elements, with glass for the rear element, which carries most of the power of the system. Of course, aspheric surfaces are just as easy to mold as spherical surfaces, but for mass production it is necessary to make a large number of identical aspheric molds, which is an extremely difficult problem. Furthermore, the hot molten plastic material tends to shrink somewhat on cooling, which may lead to the aspheric surfaces being not quite what is required to remove the lens aberrations. As this is the sole reason to use aspheric surfaces, shrinkage may well negate the intended result unless care is taken to deal with it properly.

At somewhat higher apertures, such as *f*/4.5 or *f*/3.5, the two airspaces

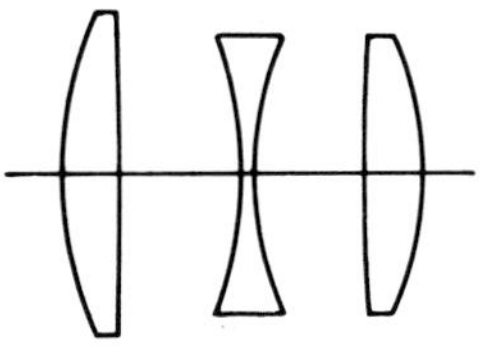

Figure 7.5. The Triplet.

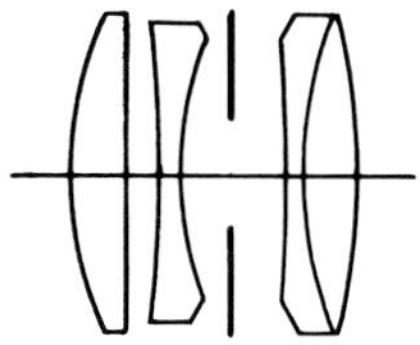

Figure 7.6. The Tessar.

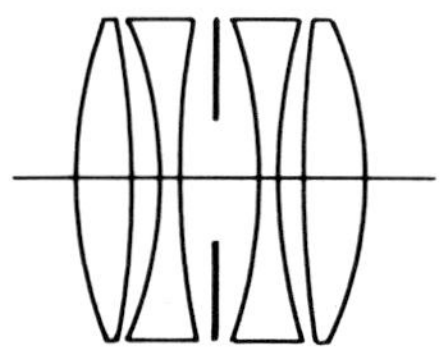

Figure 7.7. The dialyte type.

in Triplet lenses are generally made approximately equal, giving the lens a symmetrical appearance. At smaller apertures, however, the rear airspace is often made much larger than the front to provide room for the shutter and iris diaphragm. Lenses of this unsymmetrical type are often focused by a longitudinal movement of the front element alone; this greatly simplifies the construction of the camera since the shutter and the rest of the lens then remain fixed and only the front element has to be movable.

A few of the better-grade cameras are equipped with a lens of the Tessar type (Fig. 7.6). In this design the front component resembles the front of a Cooke Triplet, but behind the stop is a cemented doublet instead of a single positive element. The cemented interface in this doublet has several beneficial effects on the aberration correction, but the cost of the additional element and the required cementing operation is great enough to prevent the general adoption of Tessar-type lenses. Indeed, a Triplet made with high-index glasses is often as good as a Tessar and is considerably cheaper to manufacture.

A few lenses are still being made of the four-element dialyte type, especially for enlarging and copying applications where the lens must be usable over a range of magnifications. Dialyte lenses (Fig. 7.7) have the useful property that their aberrations do not change noticeably when the object distance is varied.

Macro Lenses

It is sometimes a great convenience to be able to focus a lens down to very close distances, especially when photographing small objects such as stamps or coins. Unfortunately, many otherwise excellent lenses do not lend themselves to this particular application, but manufacturers have found types of construction that can be so used, and these are mounted in long sliding focusing sleeves, or even with a summation thread to cause a large focusing motion with a small rotation of the focusing ring. Such lenses are called *macro lenses*. In some cases improved definition has been achieved by a differential movement of one or more of the lens elements during focusing, a lens element so moved being called a *floating lens*.

Another interesting and often useful feature is a lens in which the entire system can be raised and/or tilted, to convert an SLR camera into a kind of miniature view camera, although with somewhat limited features.

High-Aperture Lenses

For apertures greater than *f*/2.8, up to as high as *f*/1 in some cases, the so-called double-Gauss construction is generally used. Early in the nineteenth century, the mathematician C. F. Gauss became interested in geometrical optics, and among his other discoveries was a type of telescope objective consisting of a positive meniscus element of crown glass closely spaced from a negative meniscus element of flint glass. The interesting property of this objective was that its spherical aberration is constant for all wavelengths. Although this particular property is of little value to the photographer, Alvan G. Clark, in 1888, attempted to combine two Gauss telescope objectives about a central stop to form a photographic objective [Fig. 7.8(a)]. When this plan was properly worked out, it was found to lead to lenses of moderate aperture (*f*/6.3 to *f*/8) but excellent covering power. Early examples were the Ross Homocentric and Meyer Aristostigmat of 1900, and the much later Kodak Wide-field Ektars, which covered about ±35° at *f*/6.3. The Zeiss Topogon lens and the similar Bausch & Lomb Metrogon, which have been used extensively for aerial photography, were an extreme modification of this type [Fig. 7.8(b)].

In 1896, Paul Rudolph of Zeiss attempted to raise the aperture of the double-Gauss lens to *f*/4, in his symmetrical Planar lens, involving the use

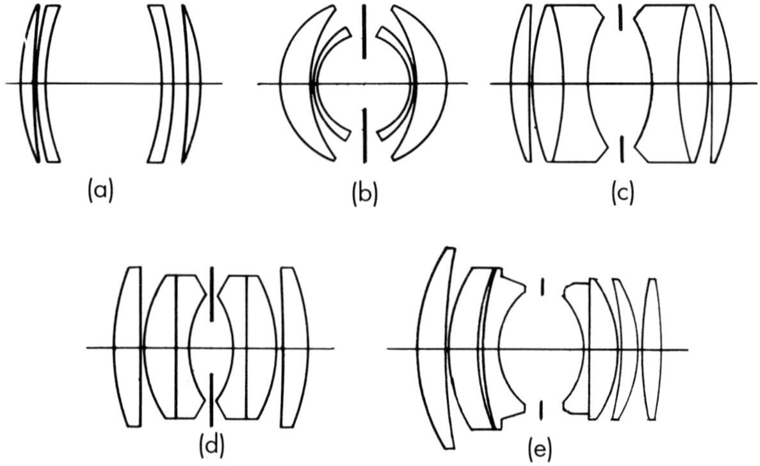

Figure 7.8. So–called Gauss-type meniscus lenses.

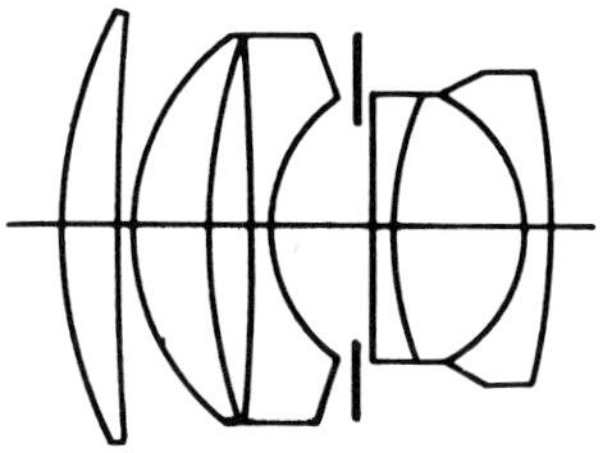

Figure 7.9. The *f*/1.5 Sonnar lens.

of two thick negative cemented doublets, as can be seen in Fig. 7.8(c). In 1920, H. W. Lee of the Taylor-Hobson company raised the aperture still further, to *f*/2, in the Opic lens [Fig. 7.8(d)] by departing from strict symmetry and changing the refractive index of the glasses. This type was used by W. Merté in the various Zeiss Biotar lenses, and by every other manufacturer since. Today, almost all the *f*/1.8 and *f*/1.4 lenses used on SLR cameras are of this general type, often slightly modified as in Fig. 7.8(e). A few manufacturers, however, favor the high-aperture Sonnar type shown in Fig. 7.9.

Some of the most recent double-Gauss lenses incorporate one or more aspheric surfaces, in an effort to raise the aperture to *f*/1.2 or *f*/1 and to eliminate small residuals of aberration that have the effect of causing a faint halo of light around light sources in night photos.

A few high-aperture narrow-field lenses of the modified Petzval type are still being manufactured, mainly as projection lenses for 8mm and 16mm movies (Fig. 7.10). These lenses have a high aperture ranging from

(a)

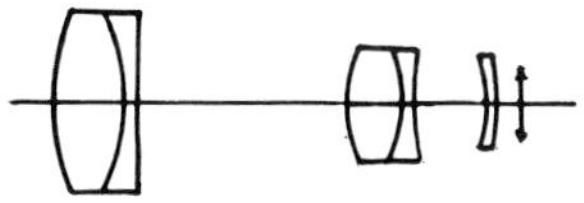

(b)

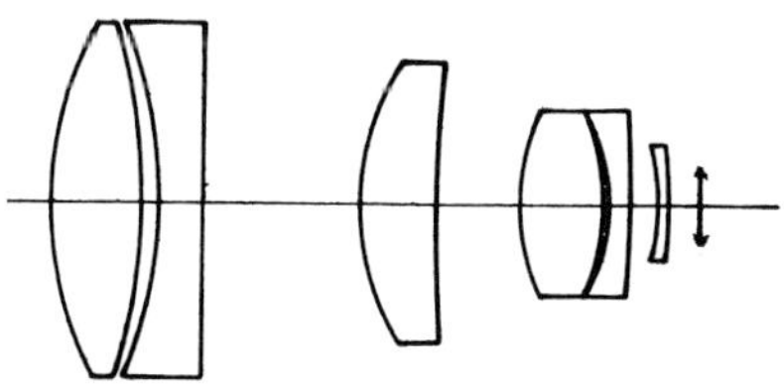

Figure 7.10. Two typical projection lenses for 8 mm film: (a) 25 mm *f*/1.6 and (b) 28 mm *f*/1.0, both drawn to the same scale.

f/1.6 to *f*/1.0, and generally consist of a pair of cemented doublets spaced apart, with a negative field flattener close to the film plane. For the highest apertures, an additional lens element is often added between the two main components.

For ordinary slide projectors, however, lenses of the Triplet type are generally employed. The focal length of the projector lens is often about twice that of the camera, and the angular fields are small, about ±7° for movie projectors and ±14° for slide projectors.

Telephoto Lenses

It was mentioned on page 49 that the telephoto lens consists of a positive front component widely separated from a negative rear component, the consequence of this arrangement being that the posterior principal plane is out in front and the total length from front vertex to film plane is shorter than the focal length (Fig. 7.11). The *telephoto effect* is the ratio of the total length to the focal length, and in most telephoto lenses its value is about 0.8. Because of the difficulty of designing good lenses of this type, telephoto lenses are used only when the total length must be kept short, as, for instance, in SLR camera lenses of focal length greater than about 150 mm. Occasionally a 135 mm telephoto lens is encountered, but there is little need to shorten the lens in this focal length, particularly as normal lenses have much smaller aberration residuals.

Some typical long-focus telephoto lenses for 35mm cameras are

Focal length	Semiangular field
400 mm	3.1°
300 mm	4.1°
200 mm	6.2°
150 mm	8.2°.

For focal lengths greater than 400 mm, it is customary to resort to catadioptric systems (see page 163). These can be regarded as extremely

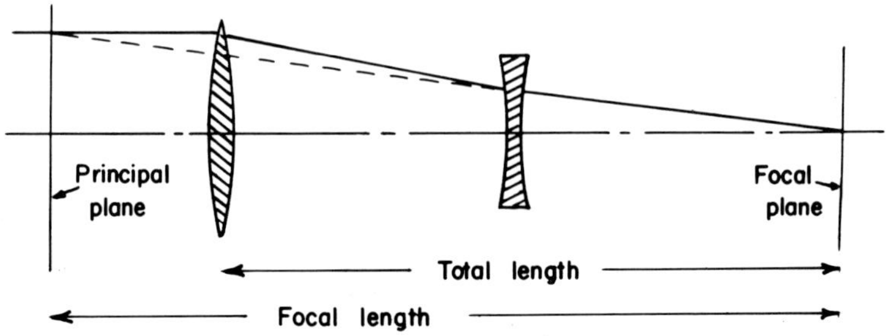

Figure 7.11. Principle of the telephoto lens.

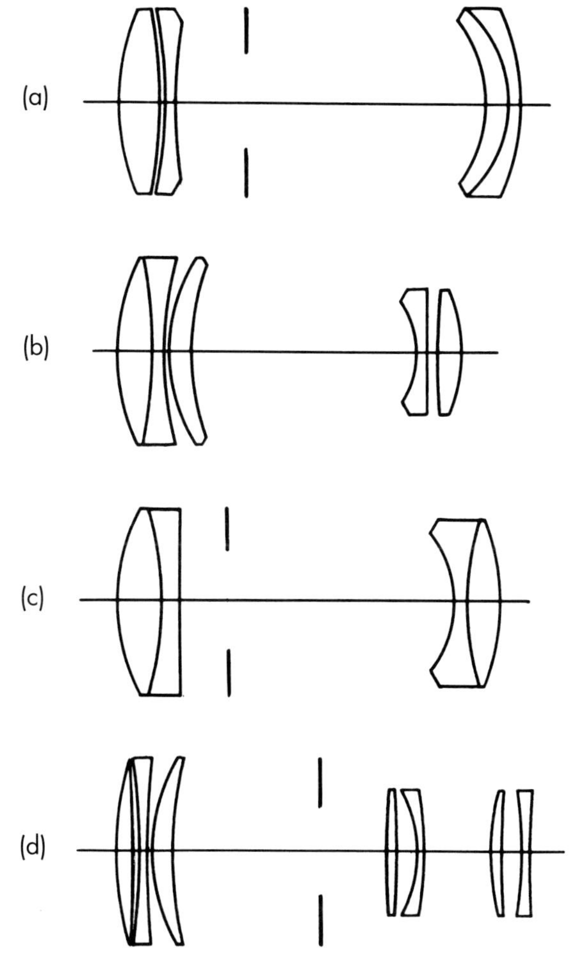

Figure 7.12. Some early telephoto lenses.

short telephotos in which mirrors have been used instead of lenses.

Telephoto lenses were formerly common when the standard film sizes were 4 × 5 and 5 × 7 inches, and they have been used extensively on aerial cameras covering a 9 × 9 inch format (Fig. 7.12). Today, aerial cameras use smaller film sizes, and telephoto lenses are therefore not nearly so important.

When telephoto lenses were first introduced, in the early 1890s, the airspace between the positive and negative components could be varied to change the focal length, making a kind of primitive zoom lens. However, the system had to be focused manually by use of the camera bellows at each

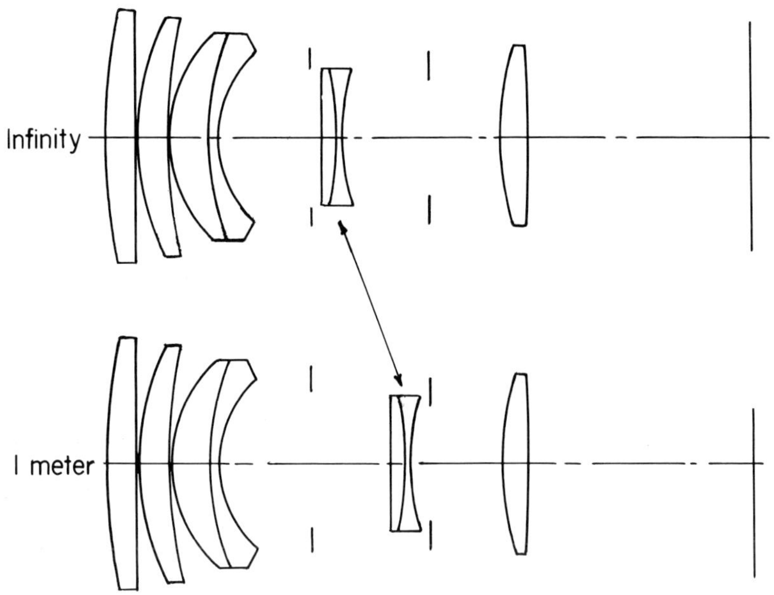

Figure 7.13. The Minolta 135 mm *f*/2.5 telephoto lens, with internal focusing.

focal length, so that they were not really zoom lenses, and the aberration correction fell off badly as the focal length was changed.

During the past few years a number of telephoto lenses have been developed in which focusing on a close object is accomplished by a movement of some internal lens elements, the large heavy front component remaining fixed. The purpose of this is to lighten the load on the tiny motor inside the camera when automatic focusing is provided.

A typical example of this is found in the current Minolta 135 mm *f*/2.8 lens for use on the Maxxum cameras (Fig. 7.13). Here the total length remains at 125 mm, with a telephoto ratio of 0.92, the front four elements and the positive rear element remaining fixed while the internal cemented doublet is moved through 13 mm toward the film when focusing from infinity down to 1 meter.

Reversed Telephoto Lenses

As was mentioned on page 50, the telephoto system can be turned around so that the negative component faces the distant object and the positive component is in the rear. The consequence of this arrangement is that the posterior principal point moves back, and the focal length becomes less than the space between the rear of the lens and the film. This is a necessary property of all short-focus lenses used on an SLR camera, because the mirror situated behind the lens requires at least 35 mm of airspace to permit it to rise and fall freely. Consequently, all SLR lenses having a focal length of 35 mm or less must be of this type, with some typical lenses as follows:

Focal length	Semiangular field	Back focus clearance
35 mm	31°	1.0 f
28 mm	38°	1.2 f
25 mm	41°	1.4 f
21 mm	46°	1.7 f
18 mm	50°	1.9 f

Sectional diagrams of two typical reversed telephoto lenses are shown in Fig. 7.14. The first reversed telephotos manufactured by the French firm of

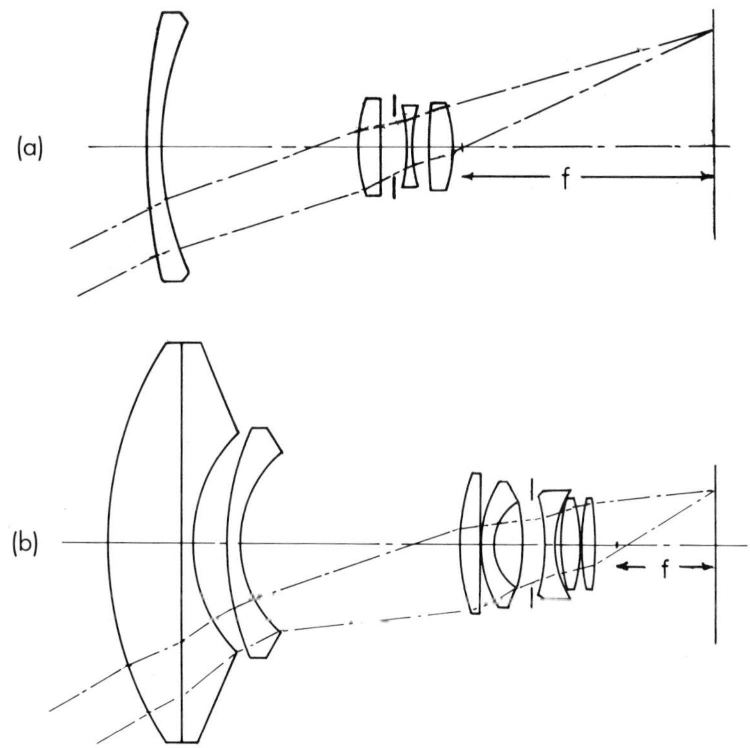

Figure 7.14. Two reversed telephoto lenses of 1953: (a) Enna Lithogon, $f/4.5$, ±25° and (b) Taylor-Hobson Speed Panchro II, $f/1.7$, ±37°.

Angénieux were named "Retrofocus," and this name is tending to become a generic term for all such lens types.

It will be seen in this figure that the iris diaphragm is small and located within the rear positive component of a reversed telephoto objective. To cover a wide angular field, the front negative component must be quite large and usually of a meniscus form; and if the field is very wide, two or more negative elements are used, often with a thin positive element to help control the distortion.

The construction of a typical modern 20 mm wide-angle lens covering an angular field of ±47° is shown in Fig. 7.15. The system is complicated, as can be seen, involving no less than ten lens elements. It is focused by a small forward movement of the rear assembly, the large two elements in front remaining fixed. The reversed telephoto type tends to be favorable for both a high aperture and a wide angular field; indeed, this type could be used for all purposes, except for its large size, complex structure, and high cost.

Reversed telephoto lenses, and indeed all lenses having a large negative meniscus element in front, tend to give increased illumination at the corners of the field as compared with more normal types of lens. If you look into the front of a reversed telephoto lens at the entrance pupil (the image of the iris diaphragm), you will see that it appears to grow in size and tilt over toward your eye as the lens is tipped. It is this increase in the size

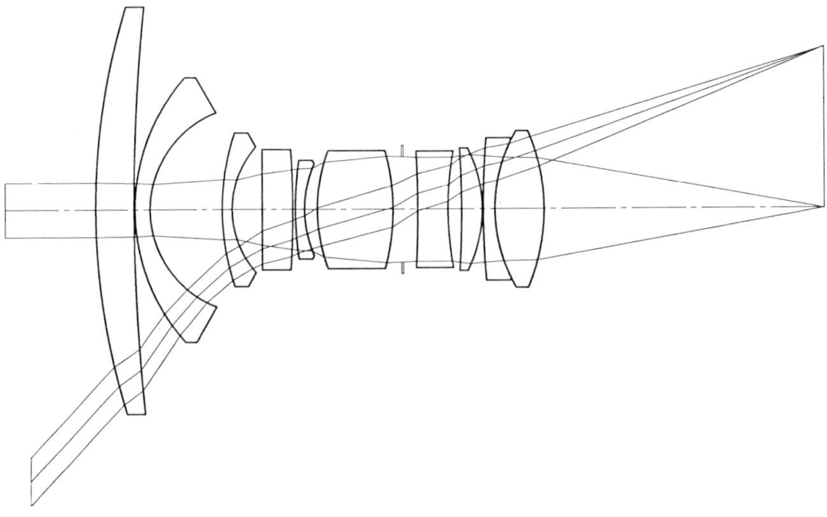

Figure 7.15. Minolta 20 mm *f*/2.8 wide–angle lens. Angular field ±47°.

of the entrance pupil at high obliquities that causes the increase in corner illumination. Lenses having a large positive element in front show the opposite effect, resulting in reduced corner illumination.

Other Wide-angle Lenses

In rangefinder cameras, and any camera that does not have a sloping mirror behind the lens, other types of wide-angle lens can be used. The best known of these is the Zeiss Biogon, based on the Roosinov principle, consisting of a strong central positive member lying between a pair of large negative menisci, somewhat like a double-ended reversed telephoto (Fig. 7.16). This lens covers a field of ±45° at *f*/4.5, with excellent aberration correction and virtually no distortion. A lens of this type, the Wild Aviogon, is now the standard lens for aerial mapping cameras because of its low distortion. Recently, Zeiss announced a simplified lens called the Hologon (Fig. 7.17), which covers a field of ±55° at *f*/8.

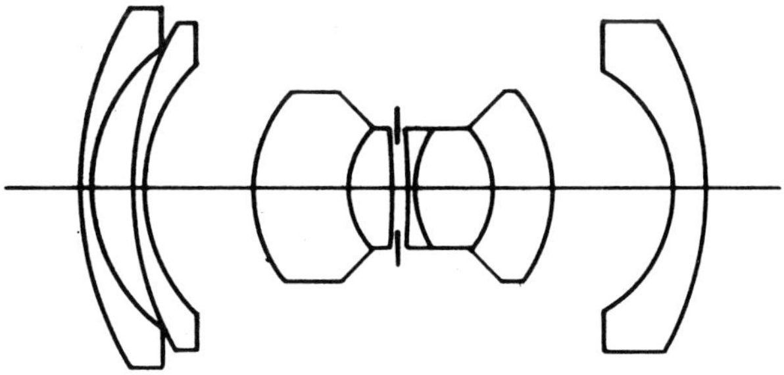

Figure 7.16. The Zeiss Biogon.

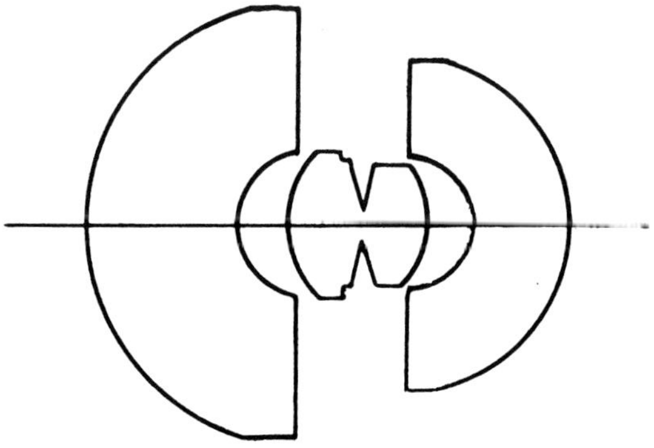

Figure 7.17. The Zeiss Hologon wide–angle lens.

Fish-eye Lenses

Distortion of the barrel type is a common defect of reversed telephoto lenses, and instead of trying to correct it, some designers have deliberately increased its magnitude in such a way that a field of 180°, a complete hemisphere, can be imaged as a finite circle on the film. If the lens is carefully designed, the radial distance h' of an image from the lens axis is directly proportional to the angle θ in the object space between the object point and the lens axis, so that $h' = f\theta$ instead of the usual law where $h' = f\tan\theta$. The construction of a typical fish-eye lens is shown in Fig. 7.18, although some designs are much more complicated than this. The large barrel distortion of a fish-eye lens leads to some peculiar effects, as shown in Fig. 7.19. In Fig. 7.19(a) the lens axis was vertical so that the zenith is in the middle of the picture and the horizon is around the outer edge, vertical objects appearing radial in the picture. In Fig. 7.19(b) the lens axis was horizontal, and here again radial lines are straight but tangential lines

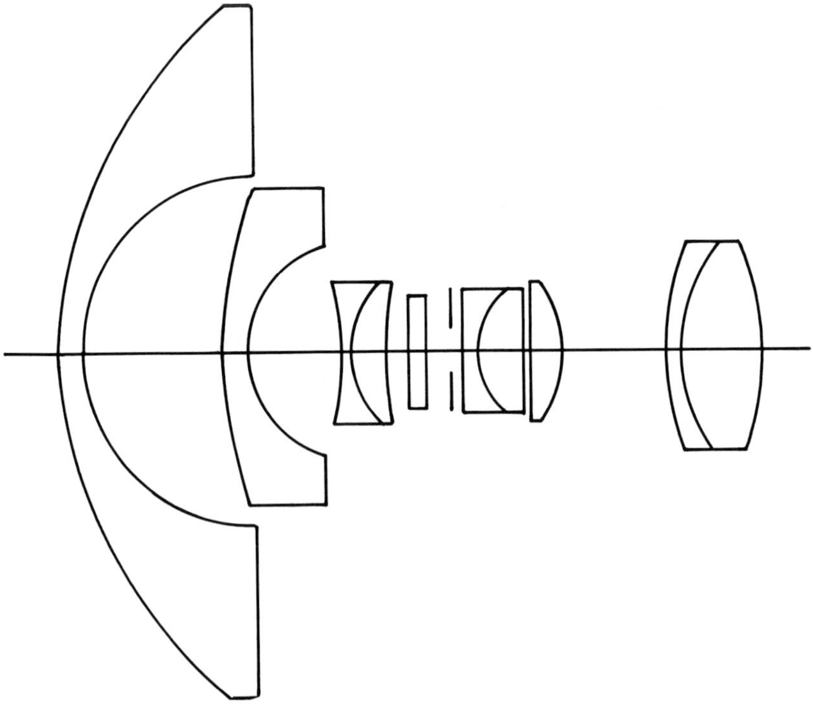

Figure 7.18. The Nikon 6 mm $f/5.6$ Fish-eye Nikkor lens (angular field ±110°).

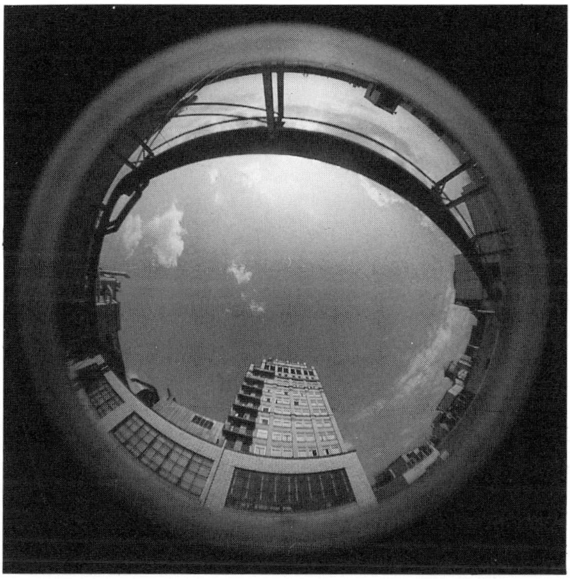

(a)

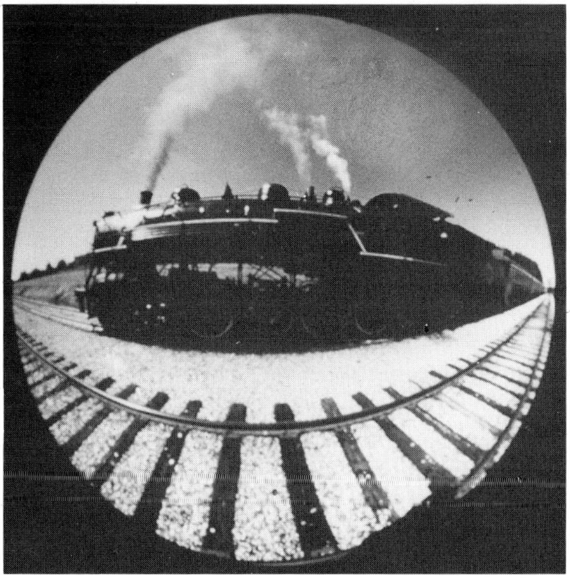

(b)

Figure 7.19. Two fish–eye photographs: (a) lens axis vertical and (b) lens axis horizontal.

are strongly curved. The focal length of a fish-eye lens is very short, leading to exaggerated perspective effects in addition to the rounding of straight lines in the outer parts of the scene. Lenses of this type have been used for technical purposes where it is desired to record an entire room in a single picture.

Zoom Lenses

Aside from a few attempts to design a so-called *varifocal* or *zoom* lens in the early 1930s, this subject may be regarded as an entirely post-war development. The first Zoomar lens for 16mm film appeared in 1945. Some professional television zoom lenses were developed around 1950, and the use of zoom lenses on 8mm movie cameras dates from 1956. Zoom lenses for 35mm still cameras followed a few years later. The rate of advance of such a difficult art during the past 35 years has been little short of phenomenal.

Zoom lenses for 8mm cameras are the easiest type to design, because the lens can be physically large compared to its focal length, and the angular field is small. At first these systems consisted of a three-lens variable power telescope mounted in front of a normal camera lens, as indicated in Fig. 7.20. In this arrangement the moving negative lens is called the *variator*, and one of the two outer positive lenses, known as the *compensator*, is moved out-and-in by a cam to maintain a fixed focal plane. The three components shown by single elements in this diagram are, of course, compound units to keep the aberrations at a minimum. One example of such a system is shown in Fig. 7.21. A prismatic beamsplitter is mounted in the parallel light between the zoom system and the fixed normal lens, to divert some of the imaging light to the viewfinder, and the photocells controlling the exposure are also placed in this part of the beam. The iris diaphragm is inside the main lens, and the rotating shutter is behind the last lens element. As shown here, the compensator lens is negative and moves in-and-out during a zoom, a fixed positive lens cemented to the beamsplitter rendering the light parallel in the cube.

Of course, it is possible to scale up an 8mm zoom lens to twice its size for use with 16mm film, but such a lens is liable to become quite large and a complete redesign is desirable. The whole subject of zoom lens design is covered in the literature.*

*See R. Kingslake, *A History of the Photographic Lens*, Academic Press, New York (1989).

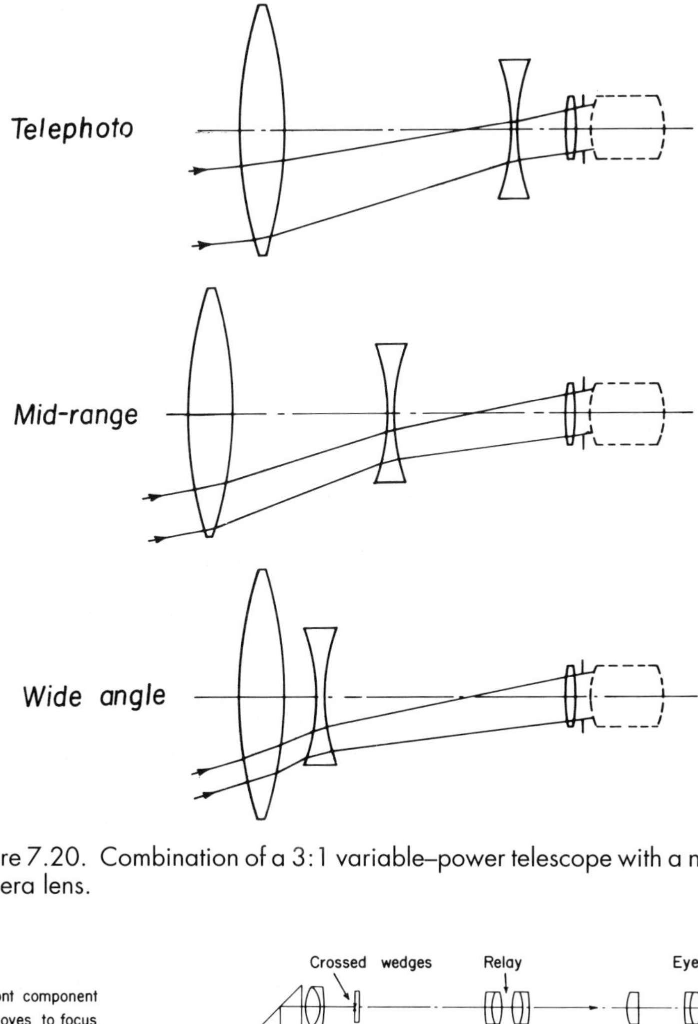

Figure 7.20. Combination of a 3:1 variable–power telescope with a normal camera lens.

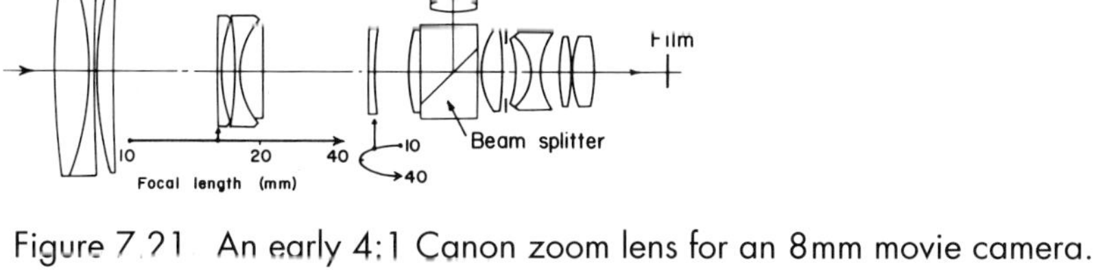

Figure 7.21 An early 4:1 Canon zoom lens for an 8mm movie camera.

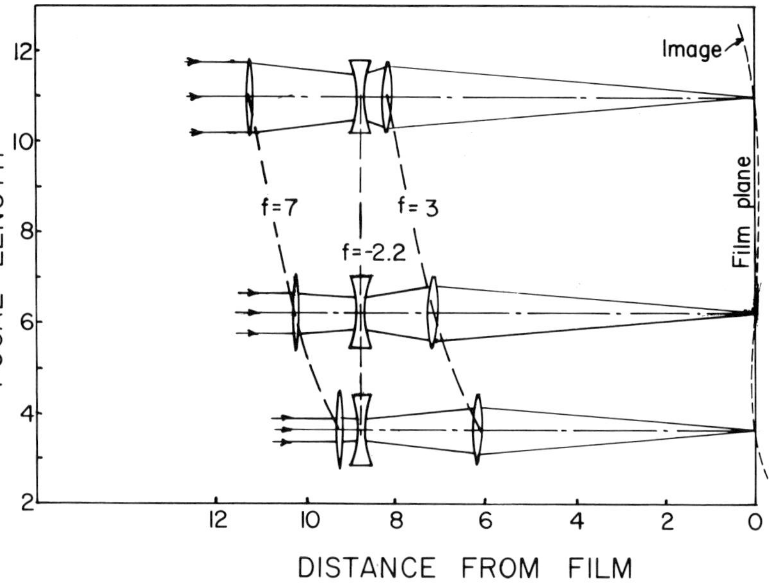

Figure 7.22. Principle of an optically compensated zoom lens.

Optical Compensation

A few zoom lenses have been constructed of the *optically compensated* type, in which two or more movable lenses are linked together with fixed lenses between or outside them (Fig. 7.22). By careful selection of the focal lengths and separations of the various lenses it is possible to design such a system so that the image moves only slightly in-and-out during a zoom, and, of course, if this movement is less than the depth of focus, the image can be regarded as being substantially stationary. As before, the iris diaphragm must be inserted behind the last moving element, and the zoom section can be mounted in front of a strong ordinary lens to reduce the focal length of the system. Optically compensated lenses were considered advantageous as they required no cam, but this turned out to be no real virtue, and the limited range of focal lengths rendered this type undesirable. Furthermore, the in-and-out image motion, though small, cannot be completely ignored, especially in lenses of high relative aperture.

In a related type, said to be *linearly compensated,* two or more components are moved either in the same or in opposite directions, through linearly related distances, relying on the small in-and-out image motion to maintain acceptable focus. This arrangement has not often been used.

Zoom Lenses for 35mm Still Cameras

When we come to design a zoom lens for a 35mm still camera, we immediately encounter a scale problem, as the diagonal of the 35mm frame is six times as great as that of the Super-8 movie frame. Obviously, we cannot scale up an 8mm zoom lens by a factor of six, as the lens would become impossibly huge. Furthermore, the angular fields required in a still camera are far wider than those of a movie camera, and consequently zoom lenses for 35mm SLR cameras must be specially designed for the purpose. The system must be compact, with an overall length comparable with the focal length, and the angular field must be quite large. Thus, it is no surprise to find that the early zoom lenses for still cameras covered only a small range of focal lengths, 2:1 or 3:1 at the most, and they had a fairly low relative aperture.

Historically, the first zoom lens for a 35mm still camera was the Voigtländer-Zoomar of 1958. This lens was optically compensated, and covered a range of focal lengths from 36 to 82 mm at *f*/2.8. Most camera manufacturers were soon making a zoom lens running from 70 to 210 mm at about *f*/4, a typical example being the Vivitar shown in Fig. 7.23. This lens is interesting as it can be converted to a macro lens by locking the zoom action at one end of its range and moving only one lens group for focusing on a close object. It is of the classical type shown in Fig. 7.20, with a moving variator B and a cam-driven compensator C. The rear group containing the diaphragm is fixed. Most zoom lenses are focused by a movement of the front lens group, so that the zoom law will not be changed for a close object. However, some zooms are focused by a movement of the whole system, which is not likely to present a problem as the user of an SLR camera

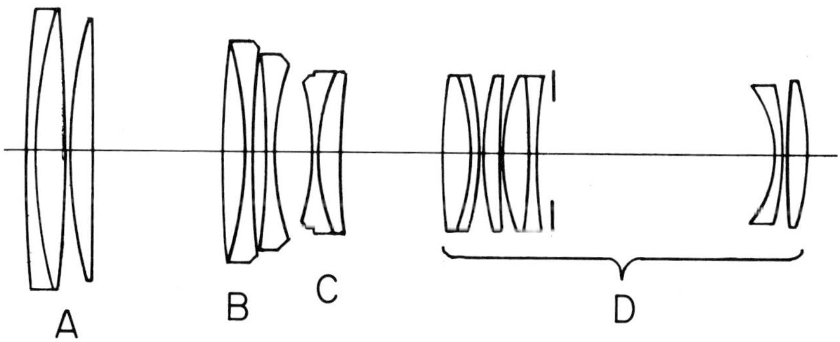

Figure 7.23. A typical zoom lens for a still camera, the Vivitar 70–210 *f*/3.5 macro zoom (shown in the 105 mm position).

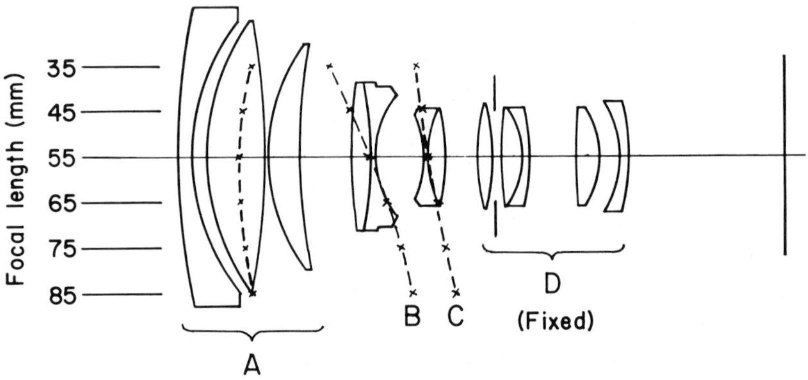

Figure 7.24. The Vivitar zoom lens, 35–85 mm at *f*/2.8.

invariably adjusts the focus manually before pressing the exposure button.

During the past few years many zoom lenses for 35mm cameras have appeared in which several lens groups are moved simultaneously during a zoom. This has permitted a much wider range of focal lengths with excellent aberration correction, and the low end of the range has dropped to 35 mm or even 28 mm in some cases. Sometimes two lens groups have been coupled together with another group in between, in the manner used in the earlier optically compensated arrangements.

One example of these more elaborate zoom lenses is shown in Fig. 7.24 representing the Vivitar 35-85 mm zoom lens at *f*/2.8. Here, two lens groups move backward at different rates while the front group moves out-and-in during a zoom. The entire system is moved forward for focusing, and the image will remain stationary during a zoom only if the object is at infinity.

A somewhat more elaborate system is the Vivitar 28-135 mm zoom at *f*/3.5-4.5 (see Fig. 7.25). Here, four lens groups are moved during a zoom, the front and rear groups being locked together and moved as a unit. Once again the entire system must be moved forward for focusing.

Another even more elaborate zoom is the Minolta 28-135 mm at *f*/4.5 shown in Fig. 7.26. Here, five lens groups are moved simultaneously, the third and fifth groups being locked together. This particular zoom has the additional property that the rear groups 3, 4, and 5 are moved together by about 1.5 mm to the rear to focus down from infinity to 1.5 meters. This feature has been included, of course, to enable the lens to be focused automatically by a small motor inside the camera. The mechanical complexity required to include both zooming and internal focusing in this way is remarkable.

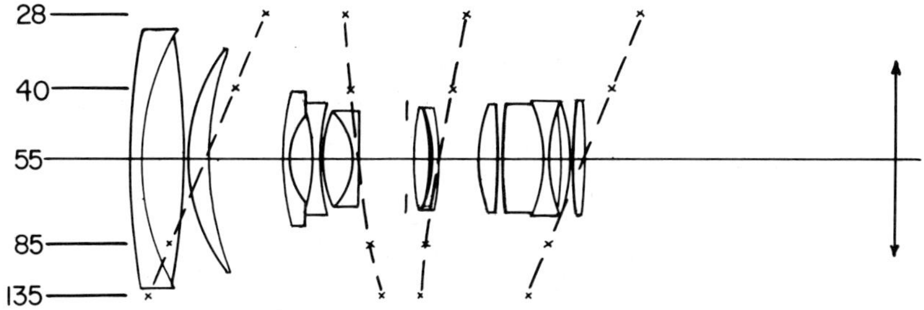

Figure 7.25. The Vivitar zoom lens, 28–135 mm at *f*/3.5 to *f*/4.5.

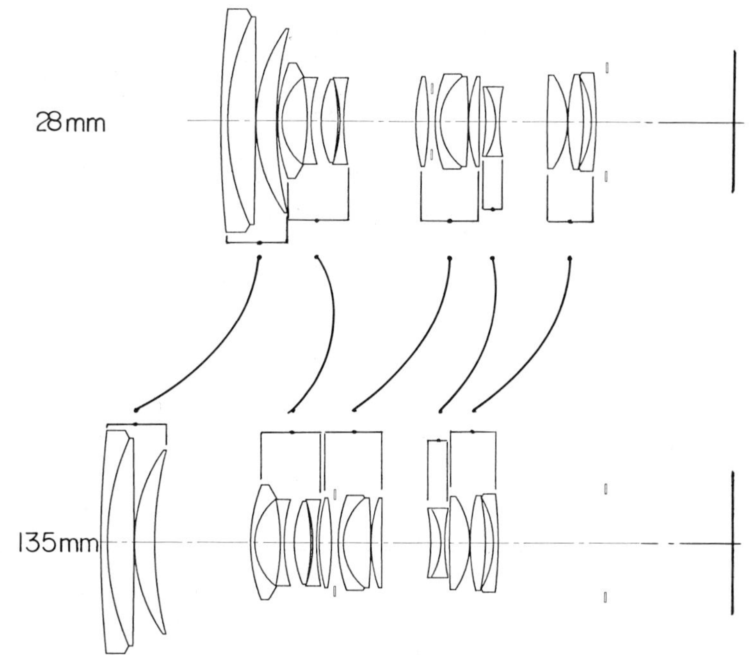

Figure 7.26 The Minolta zoom lens, 28–135 mm at *f*/4 to *f*/4.5.

Recently a new type of zoom lens has appeared, which is particularly suitable for very short focal lengths. It is really a variable reversed-telephoto system, with a negative component in front and a positive component behind, both components being moved during a zoom. The diaphragm is mounted in the rear component and also moves during a zoom, thus varying the relative aperture; however, this is not likely to be troublesome when the modern automatic exposure controls are used. The relative movements of

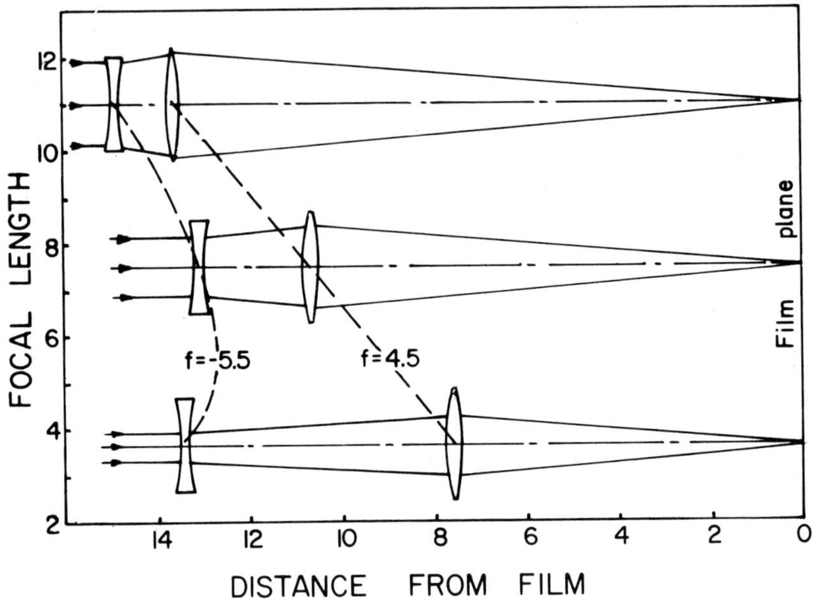

Figure 7.27. Lens movements in a two-component zoom.

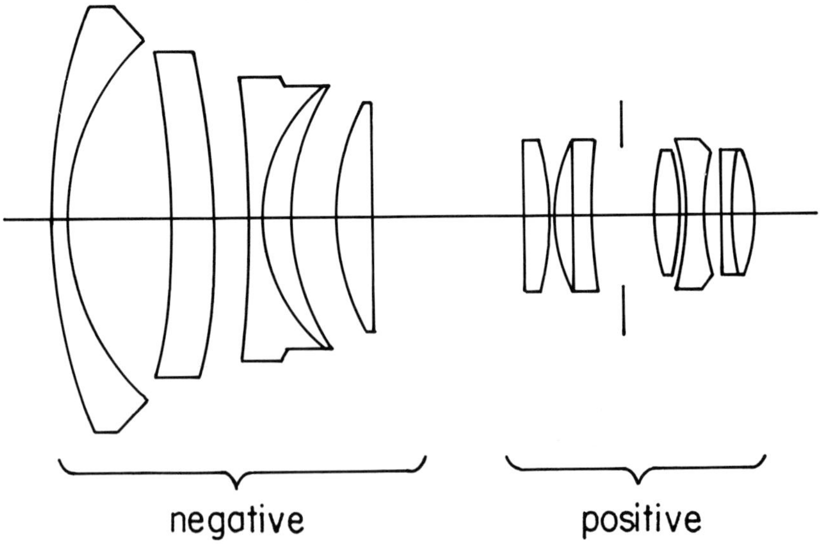

Figure 7.28. A typical two-component wide-angle zoom lens, the Canon FD 24–35 mm *f*/3.5.

the two components are indicated in Fig. 7.27, and a typical system embodying this construction is shown in Fig. 7.28.

Catadioptric Systems

The ancient terms *dioptric*, referring to lenses, and *catoptric*, referring to mirrors, are today combined in the word *catadioptric*, meaning a system that embodies both lenses and mirrors. Typical of these are the famous Schmidt camera of 1932, and the later mirror systems developed by Bouwers, Maksutov, and others. A classical Maksutov-Cassegrain catadioptric system is that embodied in the Questar telescope, which can be used as a 56-inch *f*/18 camera lens (Fig. 7.29). Other similar systems have focal lengths as long as one or two meters. A recent catadioptric system is the very compact "solid cat" manufactured by Perkin and Elmer and marketed by Vivitar (Fig. 7.30). These catadioptric systems can be regarded as extreme

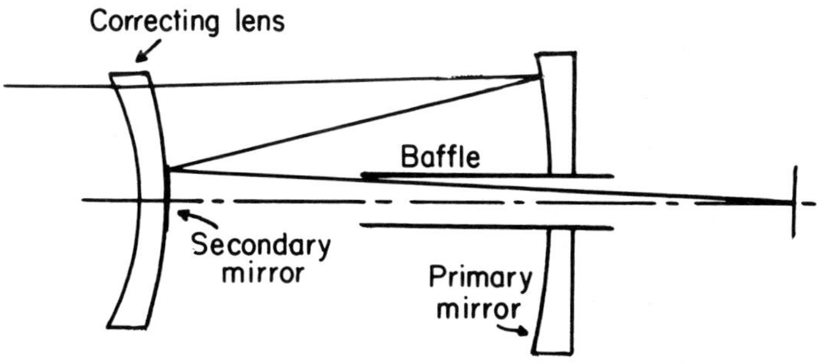

Figure 7.29. The optical system of the Questar telescope.

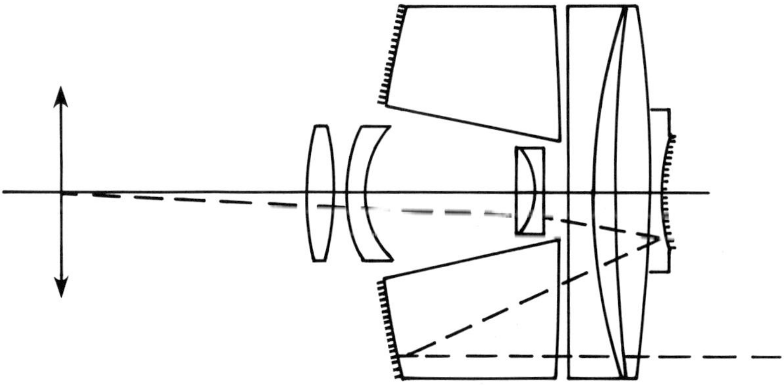
Figure 7.30. The Vivitar "Solid Cat," 600 mm at f/8.

telephotos with a total length equal to only 10 to 25% of the focal length. The mirrors in these systems are spherical, the aberration of the mirrors being corrected by the addition of suitable lenses.

It must be pointed out that the secondary mirror in a catadioptric system acts as an obstruction that blocks off some of the incident light. Also, it is generally necessary to provide baffles to prevent light from proceeding directly to the film without being reflected by the mirrors. One such baffle is indicated in Fig. 7.29.

Obsolete Lens Types

A number of lens types that used to be popular are now obsolete for various reasons. This could be because they are unduly expensive to manufacture

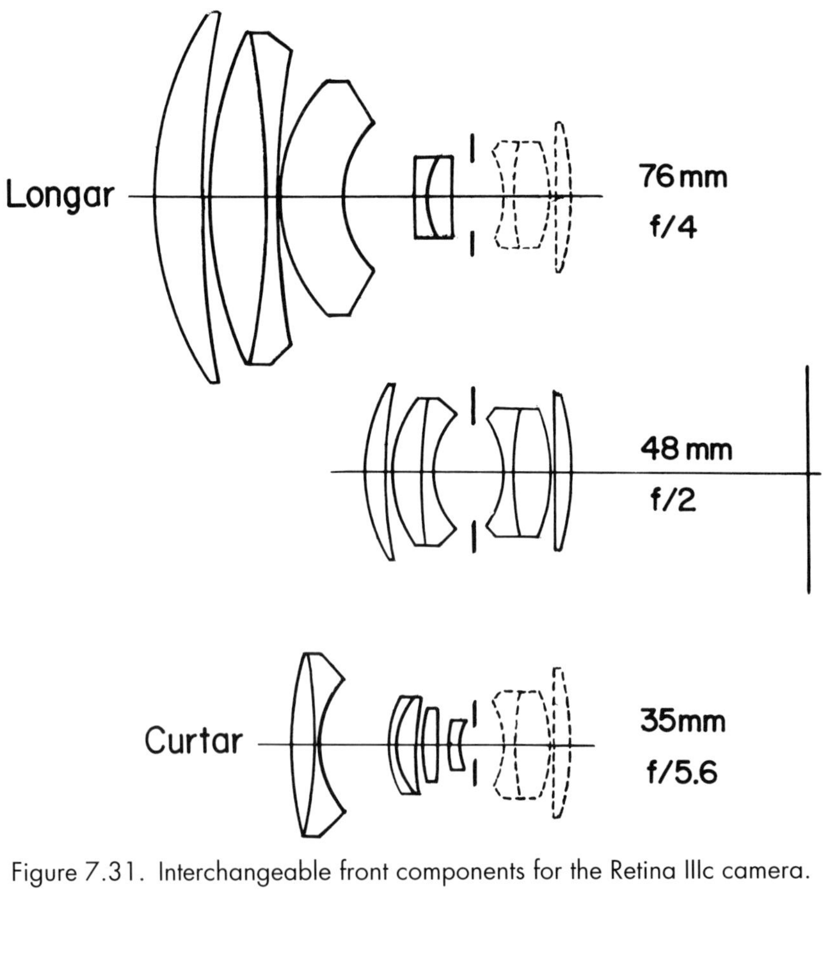

Figure 7.31. Interchangeable front components for the Retina IIIc camera.

as compared with simpler types that perform as well, or it could be because they would not cover as wide a field as more modern lens types. The list includes such well-known lenses as the Petzval Portrait lens, the Rapid Rectilinear, the Protar, Dagor, Heliar, and so on. Some of the early multielement symmetrical double objectives had the advantage of being "convertible"; that is, either half could be used alone behind the stop, or any pair of components having the same or different focal lengths could be combined together to give a focal length shorter than that of either component. Thus, two components would yield three focal lengths, so that three components with a single shutter would give the possibility of six different focal lengths.

A more recent (1955) convertible type was the short-lived series by Schneider for the Retina IIIc camera (Fig. 7.31). This series comprised the basic 48 mm *f*/2 Xenon, and the two interchangeable front components called, respectively, the "Longar," which produced a 76 mm *f*/4 system when combined with the rear component of the Xenon, and the "Curtar," which similarly produced a wide-angle lens having a focal length of 35 mm at *f*/5.6. The advantage of this arrangement was that it permitted a regular between-lens shutter to be used, but the components were complex and expensive, and in spite of their low relative apertures the definition was not as good as that of a normal lens of the same aperture and focal length. A similar system was marketed by Rodenstock. Later cameras using between-lens shutters were equipped with specially designed lenses that could be used immediately in front of the shutter. This arrangement has the advantage that a film magazine such as the 126 or 110 cartridge could be used in a reflex camera with the simple flash synchronization possibilities of a between-lens type of shutter.

Chapter 8

Lens Attachments

There are many types of attachment that may be used in front of the lens on a camera for special purposes, and indeed many camera lenses are provided with a standard mounting thread for the sole purpose of adding a hood, a filter, or some other attachment. The various common types of attachment will be described in the present chapter, with some indication in each case as to their application.

Hoods

A lens hood, or sunshade, is the simplest possible attachment that can be used in front of a lens. The purpose is, of course, to prevent direct sunlight from entering the lens at such an angle that it would fall upon the inside of the camera body or bellows, whence it might be reflected to the film and cause an unwanted flare of light.

It is seldom realized that in order to be effective, a lens hood must be surprisingly long, as can be seen in Fig. 8.1. The case shown there corresponds to a camera facing in the general direction of the sun, which is assumed to be 20° or so beyond the limit of the camera field. The minimum hood size to protect the camera in this case is indicated in the diagram, and it will be seen that the hood must be almost as large as the camera itself to be effective. A short, wide hood is almost useless in a case like this. A very good rule is that the hood should be adequate to prevent direct sunlight from falling on the lens at all; you will then be certain that no flare will arise from this source.

In some cine lenses, especially those of short focal length, the deep mount provides an excellent hood automatically; and on professional

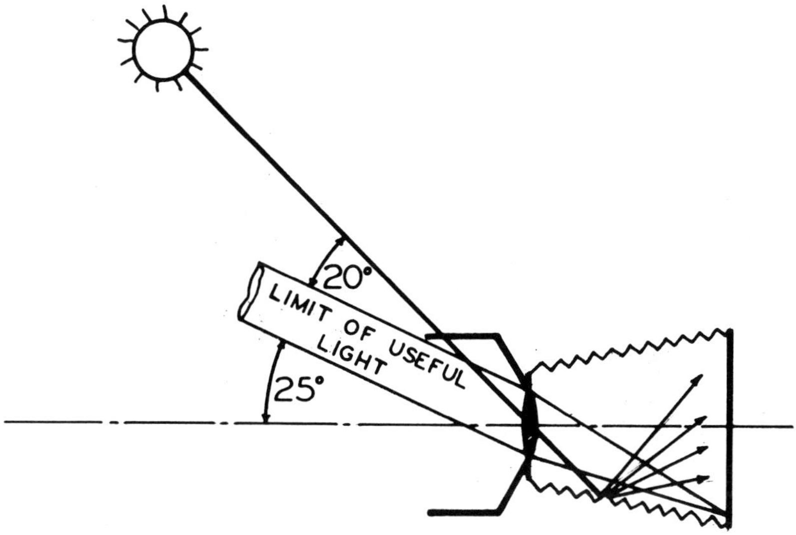

Figure 8.1. The action of a lens hood.

motion-picture cameras a large rectangular hood known as a *matte box* is often provided on the camera itself.

Filters

For many photographic purposes, a *filter* must be used. This is a piece of transparent material, gelatin, plastic, or glass, which transmits a limited range of wavelengths of light. Thus, an infrared filter transmits only light waves beyond the red end of the visible spectrum, i.e., light having a wavelength longer than about 0.75 or 0.80 μm. It is not our intention to discuss the spectral properties of filters here, since that subject is covered in books on photography, but there are a few purely optical properties of filters that must be considered.

Gelatin filters, in thicknesses of 1, 2, or 3 layers, usually do not significantly affect the definition given by a lens, provided they are clean and flat. Thumb prints on filters are as serious as thumb prints on lenses, for the same reason, and fortunately they are just as easily removed now that gelatin filters are supplied lacquered. Gelatin filters cemented between glass plates are generally satisfactory when used singly, and sometimes two may be used in succession, especially with lenses of short focus. Kodak filters in "A" glass, which is thick and very carefully optically polished, are of course harmless in all cases.

It is easy to argue that solid glass filters are likely to be better than cemented filters, because in the latter case the manufacturer has to polish four surfaces instead of only two, and the pressure applied during the cementing operation may warp the thin sheets of glass, but in practice there is no discernible difference between the two types. As a matter of fact, there are only a very few colors that can be achieved in solid glass at all, and if a wide variety of spectral transmissions is required, it is essential to use dyed gelatin or plastic.

If a glass or cemented filter is inserted into a *converging* beam of light, it shifts the image away from the lens by an amount equal to about one-third of the thickness of the filter (Fig. 8.2). Moreover, if the lens is of very high aperture (*f*/2 or higher), the presence of the filter in the imaging beam may cause a slight deterioration of the image due to the introduction of aberrations. However, this latter effect is not large and can generally be neglected.

If a filter is inserted into a diverging beam, it makes the beam appear to come from a different point, the apparent shift being again equal to about one-third of the thickness of the filter. In parallel light, of course, the presence of a filter makes no difference whatever.

A glass or cemented filter might be slightly wedged, so that it produces a small lateral displacement of an image when inserted into a beam of light. This effect is greatest when the filter is close to the lens, and zero when close to the object or image; however, it is usually so small that it can be neglected entirely. For all of these reasons, the photographer is cautioned never to insert a glass

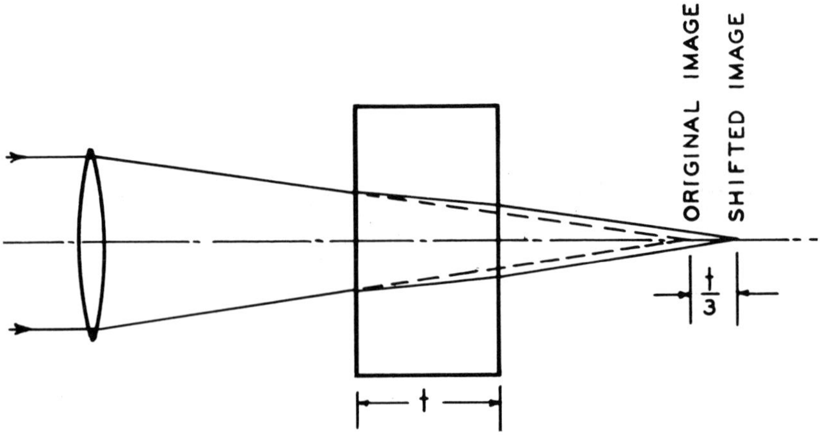

Figure 8.2. Shift of an image caused by insertion of a thick glass filter.

filter into a light beam for part of an exposure and remove it for the remainder.

Polarizing Filters

One type of filter must be especially mentioned, namely, the *polarizing filter*. There are now available several types of sheet polarizing material, which polarize the light passing through them so that all the vibrations then lie in one direction (page 60). Polarized light has unusual reflecting properties that make it of great interest to photographers. For example, as was explained in Chapter 3, light reflected from glass surfaces will be strongly polarized to an extent depending on the precise angle of incidence, and this light can therefore be extinguished by means of a polarizing filter mounted in front of the camera lens and rotated to the correct orientation (Fig. 8.3). The desired position is generally found by looking directly through the filter before attaching it to the camera and suitably rotating it in its holder while watching the scene to be photographed. A polarizing filter has little effect on reflections from polished metal surfaces, which are practically nonpolarized. As the light from a low sky, especially at right angles to the direction of the sun, is partially polarized, this too will vary in brightness as the polarizer is rotated. Thus, by mounting a polarizing filter in front of a camera lens, many unusual effects can be produced, especially those involving reflections. This is the only practical means of controlling the sky brightness in color photography, where the familiar yellow or red filters used for such purposes in black-and-white photography obviously cannot be used.

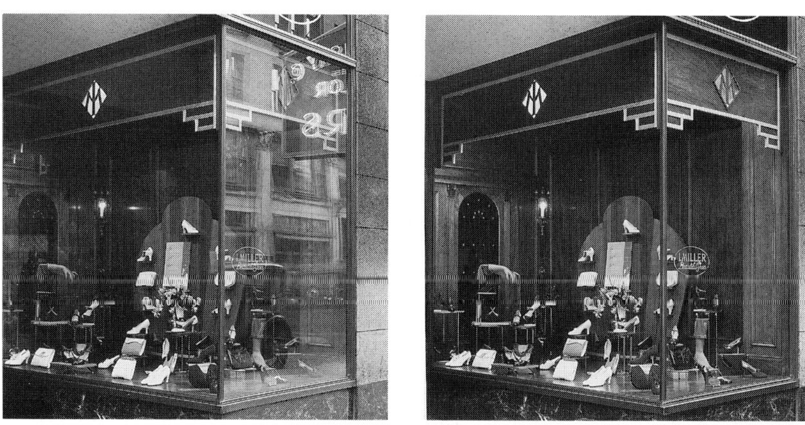

Figure 8.3. Illustrating the ability of a polarizing filter to reduce oblique reflections from glass.

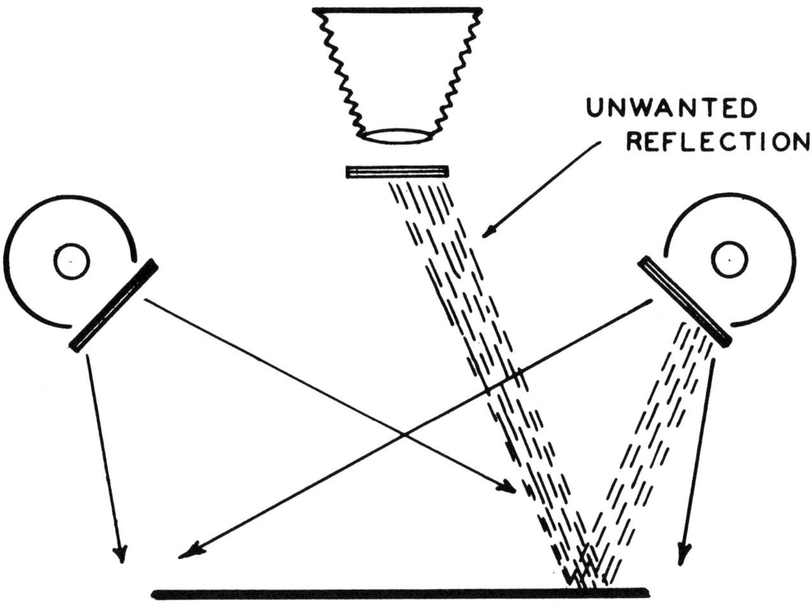

Figure 8.4. Procedure for copying drawings under a glass plate, using a polarizing filter in front of both lamps and lens.

It will be noticed, too, that when two sheets of polarizing material are "crossed," one in front of the other, almost no light is transmitted. This property is of great value in copying oil paintings under glass, for instance, or objects in a showcase. For this purpose one polarizing filter may be placed over the lamp and another over the lens, the second one being rotated to such a position as to extinguish unwanted light from the lamp (Fig. 8.4). The only light that will be recorded is that which has been depolarized by the diffusing properties of the objects themselves. This light may, however, be very faint and require a long exposure. Since polarizers and color filters act in quite independent ways, it is entirely possible to use a polarizing screen and a color filter in combination in front of a lens.

Nonuniform Filters

For some special purposes, filters that are not uniform over their entire area have occasionally been made. A typical example is the *sky filter*, consisting of a disk of gelatin dyed yellow at the top and clear beneath, which is mounted a short distance ahead of the lens [Fig. 8.5(a)]. The purpose is to make clouds stand out without any corresponding filtering action being applied to the foreground objects. To be effective, this type of filter should

(a)

(b)

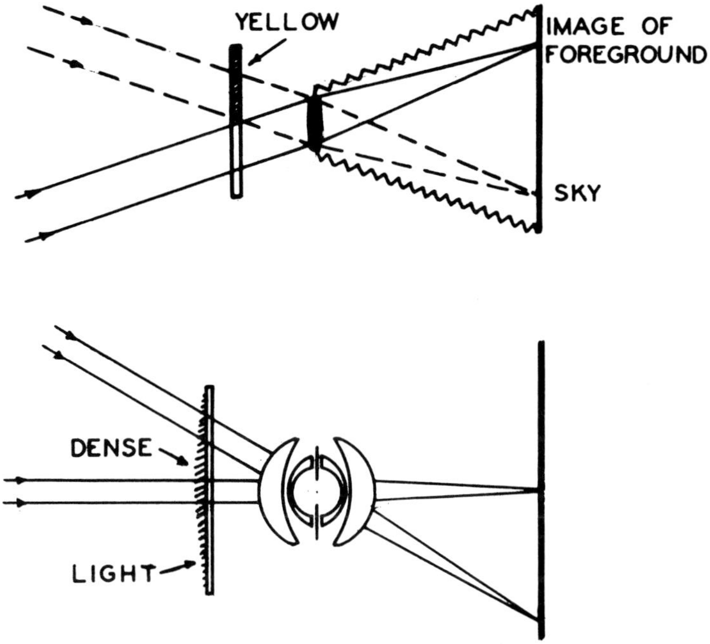

Figure 8.5. (a) Principle of the sky filter and (b) a vignetting compensating filter.

be used only with an orthochromatic emulsion on a camera of low relative aperture.

Another type of nonuniform filter is that used to equalize the illumination over the field of a wide-angle lens such as an aerial camera lens. In this, a neutral density is distributed over a glass plate with a radial distribution such as to compensate the vignetting and $\cos^4$ factors in the lens [Fig. 8.5(b)]. The filter may be mounted at any suitable position except close to the lens itself. One proposal is to expose a photographic plate slightly in the image plane of the camera while pointing at the sky, develop and fix, and then allow this plate to come into contact with the actual film used in the camera. The density distribution developed on the plate will then automatically compensate the nonuniformity of illumination present in the camera. Another proposal is to make the filter out of a thin planoconvex lens of dense neutral glass with the curved side cemented to the curved face of a similar planoconcave lens of clear glass to make a parallel plate. The density distribution in this combination will approximately compensate the vignetting of most ordinary lenses.

Use of Filters in Black-and-White Printing

An interesting application of colored filters is found in the making of enlarged prints from a black-and-white negative. Such printing papers as Kodak Polycontrast, Ilford Multigrade, or Dupont Varigam act as if they contained two emulsions, either mixed or in separate layers. One of these

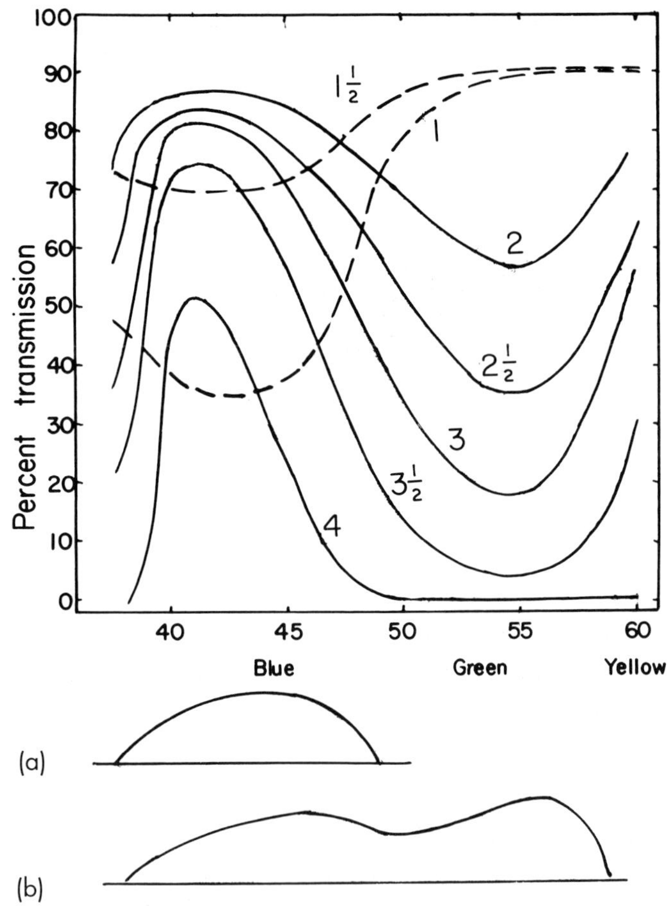

Figure 8.6. The percentage transmission of Kodak Polycontrast filters, plotted against wavelength. When no filter is used, the result is a print contrast between that of a No. $1^1/_2$ and a No. 2 paper. The upper dashed graphs represent yellow (minus-blue) filters for a low-contrast print, while the solid curves are for magenta (minus-green) filters to give a high-contrast print, the filter numbers indicating the equivalent paper grades. The two lower graphs indicate the spectral response of (a) a color-blind printing paper, and (b) an orthochromatic paper.

emulsions is blue sensitive, quite rapid, and of high contrast. The other emulsion is orthochromatic, i.e., sensitive to both blue and green light, relatively slow, and of low contrast.

With no filter in the beam, such papers have about the contrast of a normal No. 2 paper, but by the use of appropriate filters the operator can vary the contrast over a wide range. When using a yellow filter, which absorbs blue light, the result is a print of low contrast, while with a magenta filter, which absorbs green light, the result is a print of high contrast. The paper manufacturers supply sets of filters containing varying degrees of minus-blue and minus-green dyes by which the user can obtain a range of print contrasts, from the equivalent of No. 0 or No. 1 paper up to that of a No. 4 or No. 5 paper. The Kodak Polycontrast filters have the transmissions indicated in Fig. 8.6. It is obvious how the possibility of such a range of contrasts in a single packet of paper can effect a considerable saving of storage space and expense to both the retailer and the user.

Supplementary Lenses

Positive Supplementary Lenses

The use of a lens of about 4 or 5 feet focal length as a supplementary "portrait attachment," to enable a fixed-focus camera to be focused on close objects, is a very old device. The name, however, is misleading since portraits taken with this attachment when viewed without enlargement actually exhibit rather poor perspective, and it is now becoming customary to call them "close-up" attachments instead. The Kodak Close-up Attachment had a focal length of 52$^1/_2$ inches, but as box cameras are usually focused initially at about 20 feet, the correct object distance when the attachment was used worked out to be only about 42 inches.

Positive supplementary lenses ("close-up lenses") of higher power are supplied by many manufacturers. They are positive lenses, of a weak meniscus form, to enable the camera to be focused on objects at distances that are closer than the near end of the normal focusing scale. The power of these lenses is commonly expressed in diopters, a diopter being the power of a lens having a focal length of 1 meter (39.4 inches). A 2-diopter lens has a focal length of 0.5 meter, and so on (see page 30). The focus scale of the camera becomes much shortened by the use of such a lens, as is shown in the table below. This is a universal table, applicable to any camera regardless of focal length.

It will be noticed that when the camera is set at infinity, the distance of the object from the supplementary lens is just equal to the focal length of that lens. In this case a beam of parallel light exists between the attachment and the camera lens. In all other cases, the light emerging from the supplementary lens will be a diverging beam.

Object Distances with Supplementary Lenses

Camera focusing scale set at	0.75 diopter close-up attachment	+1 diopter lens	+2 diopter lens	+3 diopter lens
∞	52.5 inches	39.4 inches	19.7 inches	13.1 inches
50 feet	48.3	36.9	19.0	12.8
25	44.7	34.8	18.5	12.6
15	40.7	32.3	17.7	12.2
10	36.5	29.6	16.9	11.8
8	33.9	27.9	16.3	11.5
6	30.4	25.5	15.5	11.1
5	28.0	23.8	14.8	10.8
4	25.1	21.6	14.0	10.3
3½	23.3	20.3	13.4	10.0
3	21.4	18.8	12.7	9.6
2	16.5	14.9	10.8	8.5

A supplementary lens is generally made to a meniscus shape, but the exact form must be somewhat of a compromise, for if it is made too flat, the spherical aberration is improved but the field is somewhat inward-curved, and if it is made strongly meniscus, the field becomes flat but the central definition is not good except at very low apertures. For the shape actually employed, it is strongly advisable never to use the camera lens full open but to stop it down to at least *f*/8 or *f*/11. At these apertures, the definition obtainable with such a lens is excellent, but it must be very carefully focused as the depth of field is surprisingly small.

The *F*-number markings on the camera are unaffected by the use of a positive attachment. If the camera is focused by its front element, the *F*-number markings are correct at all positions of the focus scale and with all attachments, because in this type of lens, the cone angle subtended by the lens aperture at the film remains always constant. If the camera is focused by moving the entire lens forward, then the effective *F*-number will be given by

$$(\text{marked } F\text{-number}) \times \frac{(\text{focal length}) + (\text{focusing shift of lens})}{(\text{focal length})}$$

in the usual way. This latter rule applies also when an attachment lens is added.

Field of View with Supplementary Lenses

If the camera is focused by its front element, the angular dimensions of the field of view remain constant no matter how the camera is focused and what supplementary lenses are used. Thus, we have the rule that

$$\frac{\text{Field dimension}}{\text{Corresponding film dimension}} = \frac{\text{Object distance}}{\text{Focal length of camera lens}}$$

As an example, we may suppose that we are using a +2 diopter lens attached to a 2-inch camera lens that is focused by its front element to 10 feet. The film dimensions are $1 \times 1^1/_2$ inches. From the table on page 174 we see that the object distance must be about 17 inches, and hence the field size will be magnified by $17/2 = 8^1/_2$ times to $8^1/_2$ by $12^1/_2$ inches. Corresponding values for other object distances and diopter lenses are shown in the chart (Fig. 8.7). As a matter of fact, that chart applies also to the case of a $2^1/_4$-

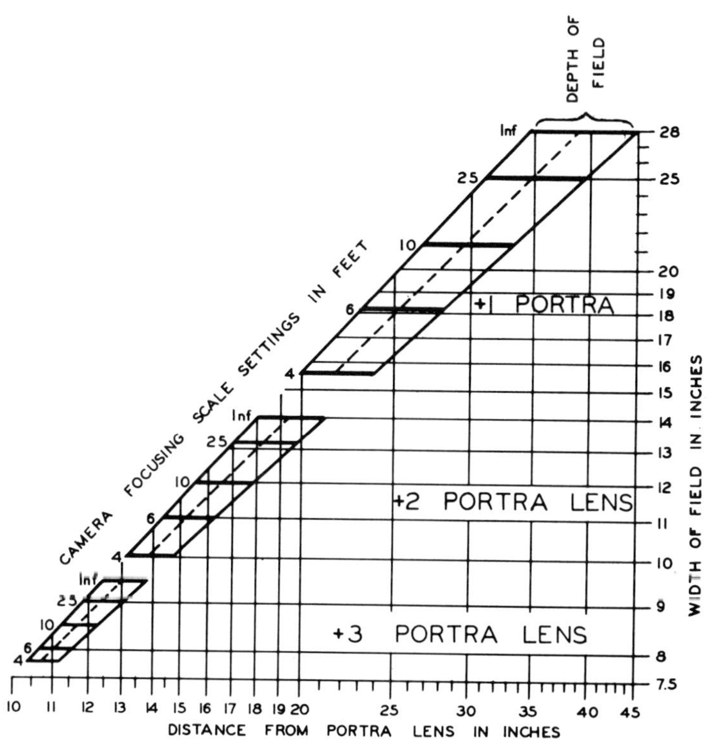

Figure 8.7. Object distance, field size, and depth of field with positive diopter lenses. (This chart applies to a 50 mm *f*/8 lens with 35mm film, or an 80 mm *f*/12 lens with a $2^1/_4 \times 2^1/_4$-inch film).

inch square picture taken with an 80 mm lens, because the ratio of $2^1/_4$ inches to 80 mm is exactly the same as the ratio of 36 mm to 50 mm, namely, 0.72. A similar chart could be easily plotted for any combination of film size and lens focus.

If the camera lens is focused as a unit, the angular field of the camera becomes somewhat reduced as the lens is focused out to near distances, and a formula is necessary to compute the fields when diopter lenses are added. If f is the focal length of the camera lens and s is the focused distance of the camera lens, the distance s' from lens node to film is given by

$$s' = \frac{sf}{s-f} . \tag{8.1}$$

The proportional rule now becomes

$$\frac{\text{Field dimension}}{\text{Corresponding film dimension}} = \frac{\text{Object distance}}{\text{Image distance } s' \text{ in camera}} .$$

Of course, no calculations of this kind are necessary with an SLR camera, as the limits of the field of view are clearly visible in the viewfinder.

Depth of Field with Supplementary Lenses

Provided the camera is focused by its front element, the diameter of the entrance pupil d is substantially independent of the focusing adjustment or the use of a supplementary lens. Thus, at any F-number N, the hyperfocal distance can be found. It is here assumed to be equal to 1000 times the lens aperture, or

$$h = 1000d = 1000 \frac{f}{N} . \tag{8.2}$$

The regular depth of field equations given in Eq. (5.4) of Chapter 5 may be used, namely,

$$D_1 = \frac{s^2}{h-s} , \quad \text{and} \quad D_2 = \frac{s^2}{h+s} . \tag{8.3}$$

As an example, we will compute the depth for the case of a 50 mm $f/8$ lens focused on 10 feet, with a +2 diopter lens in addition. The object distance is found from the table to be about 17 inches, which becomes our quantity s for the formula. We then calculate

$$h = 1000\left(\frac{50}{8}\right) = 6250 \text{ mm} = 246 \text{ inches} .$$

Then

$$D_1 = \frac{17^2}{246 - 17} = 1.26 \text{ inches}$$

$$D_2 = \frac{17^2}{246 + 17} = 1.10 \text{ inches}$$

These data are included in Fig. 8.7.

The same depth limits shown on the chart can also be applied to any other lens having the same entrance pupil diameter, namely, 50/8 = 6.25 mm. Examples are an 80 mm lens at $f/12.8$ or 105 mm lens at $f/16.8$.

Focal Frames

Since the depth of field of a camera fitted with a supplementary lens is extremely small, it is very helpful to construct a *focal frame* to locate the object accurately, both as to its longitudinal position and the area covered by the film. Such a frame may be extemporized with a bent wire and a piece of wood (see Fig. 8.8). The precise focal setting can be determined by

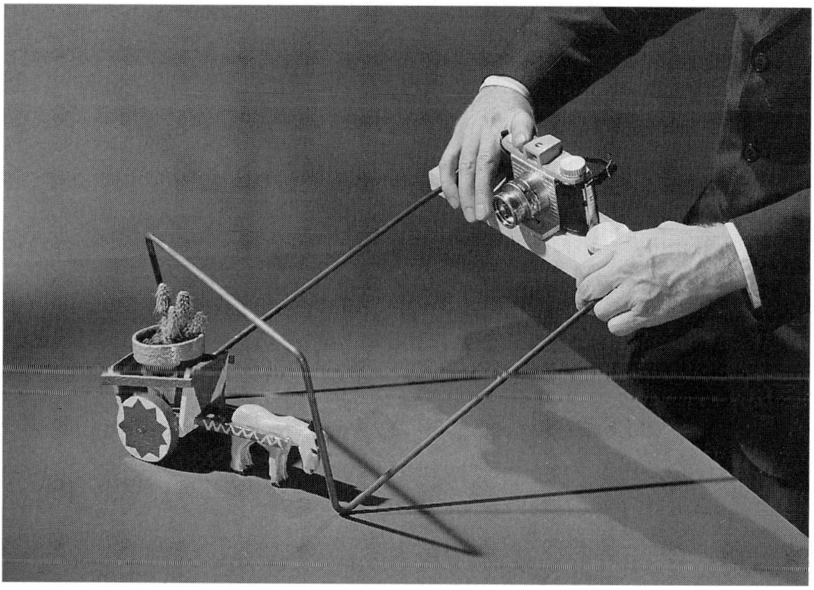

Figure 8.8. A simple homemade focal frame.

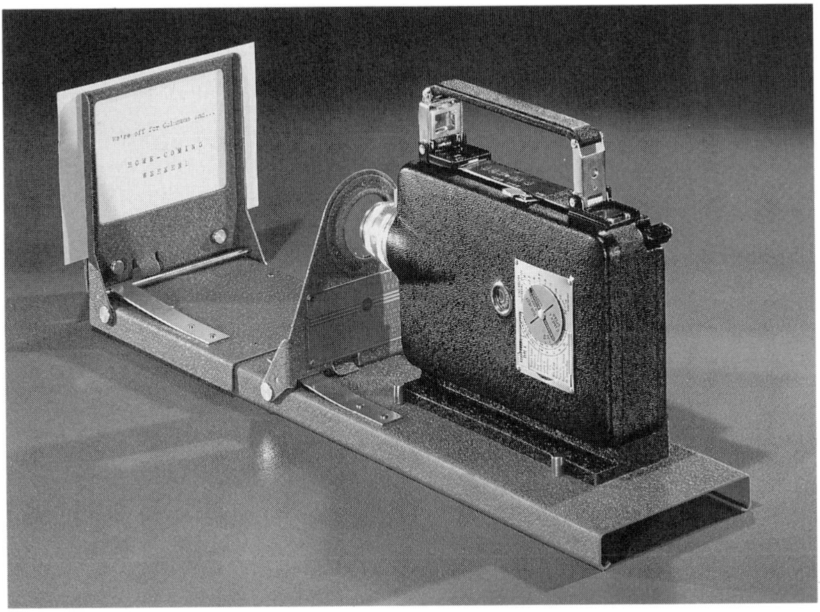

Figure 8.9. A motion-picture titler. Note the supplementary lens mounted in front of the camera.

photographing a ruler set obliquely in the frame and noting which scale division is in sharpest focus on the developed film. This scale division will then represent the best focal plane of the lens, and the wire frame can be placed at that precise position. No tripod is needed with a close-up system such as this, and it is actually quite possible to move the camera with its focal frame to an object and hold it in position while taking the photograph, provided the exposure time is short.

The familiar motion-picture titler, one example of which is shown in Fig. 8.9, is really nothing but a positive supplementary lens of 4 or 5 diopters power combined with a focal frame, the whole being mounted on a convenient support to hold the cine camera. Titles and trick devices, cartoons, etc., can be placed in the frame and photographed as desired.

Other Uses of Positive Supplementary Lenses

Since the addition of a positive lens to an existing objective shortens its focal length, we may use these supplementary lenses for that purpose. The effect is shown graphically in Fig. 8.10.

It is not generally realized that shortening the focal length of an enlarger lens permits both a higher and a lower extreme magnification to be obtained with a given film-to-easel distance. Thus, for example, the maximum and minimum film-to-easel distances of a certain enlarger were 36 and 11.75 inches, respectively, and because of the limited bellows extension the range of magnifications obtainable with a 4-inch lens lay between 6.9× and 0.96×. On adding a +3 diopter lens to the enlarging lens, the focal length was reduced to 3.28 inches, and the magnification could then be raised to 8.9× and reduced to 0.6×. It should be noticed, however, that the addition of a simple positive lens to an enlarging lens in this way introduces a small amount of lateral color, which may be great enough to render the lens unsuitable for making three-color separation negatives.

Since the addition of a positive attachment shortens the focal length of a camera lens, many photographers might expect that it will also widen the field. However, this would be true only if the field covered by the original lens were considerably less than its circle of good definition. For example, if a normal $8^1/_2$-inch lens is being used on a 4 × 5-inch camera, its normal 5 × 7-inch field is not being completely utilized. In that case the addition of a 2-diopter attachment would reduce the focus to 6 inches, which is the normal value for a 4 × 5-inch camera lens. However, it must not be expected that the definition at full aperture would be as good as if a proper 6-inch lens had been substituted for the $8^1/_2$-inch lens. The effective speed of the lens will be slightly increased because it has been moved somewhat closer to the film.

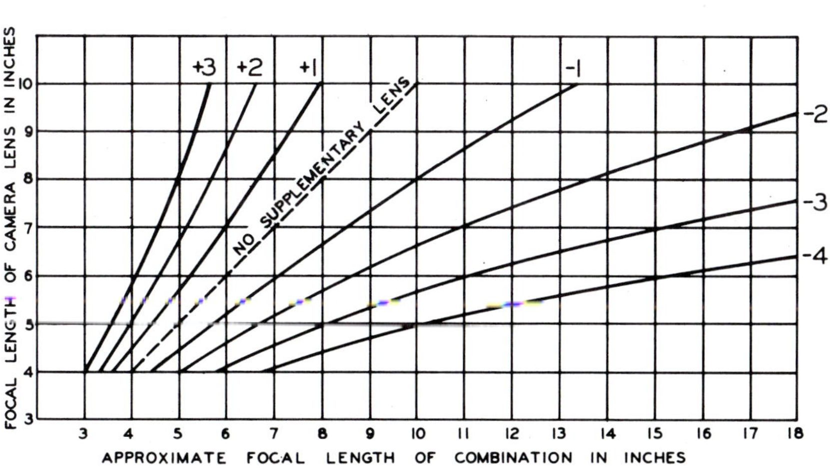

Figure 8.10. Change in focal length caused by a supplementary lens.

Negative Supplementary Lenses

A negative meniscus-shaped supplementary lens can also be used in front of a camera objective, but its action is entirely different from that of a positive lens. Adding a negative lens to a camera lens increases its focal length, thus forming an enlarged picture and reducing the angular field for a fixed subject-to-camera distance. The effective speed of the lens becomes reduced as the lens is racked forward, in the usual way. A negative attachment added to a *fixed-focus* camera cannot be used with objects at any distance whatever, and indeed such an attachment is of value only on cameras provided with a long extension of the bed. The effect of these lenses on the focal length of the camera lens is shown graphically in Fig. 8.10.

Amplifier Lenses

Early in the history of slide projectors it was discovered that the size of the projected image on a screen could be increased by the addition of a large negative lens out in front of the regular projection lens, with, of course, subsequent refocusing of the projector. The combination of a regular lens and an amplifier must have a focal length that is *shorter* than that of the main lens if it is to produce a larger image, and this result can be obtained only if the amplifier is located in front of the front focus of the main lens. A negative lens added close to a positive lens will increase its focal length, which is not what is wanted.

As with any attachment lens, the definition is bound to suffer because the main lens is fully corrected, while the added lens is only partially corrected at best. Nevertheless, most attachment lenses prove to be good enough for the intended application.

Extension Tubes and Bellows

If a camera is equipped with a removable lens, then it is far better to focus on a close object by moving the lens forward than by the use of a diopter attachment. This can be done with a *macro* lens having an exceptionally long focusing range, or by the insertion of a small bellows between the lens and the camera. As an alternative to a bellows, one may use an *extension tube* inserted between the lens and the camera body. These tubes come in multiples of 5 or 10 mm in length, and their use has a considerable effect on the object distances that can be focused. Indeed, by combining the ordinary focusing scale of the lens with one or more extension tubes, a very wide range of object distances can be covered, from infinity down to unit magnification. In fact, even a small degree of actual magnification can be obtained if a bellows is used.

To determine the lens setting for any given object distance and extension tube, the following formulas are useful:

$$P = f\left[\frac{Sf + E(S+f)}{f^2 + E(S+f)}\right] \quad \text{and} \quad S = f\left[\frac{Pf - E(P-f)}{f^2 - E(P-f)}\right], \qquad (8.4)$$

where P represents the actual distance of the object, S is the reading on the focusing scale of the lens, E is the thickness of the extension tube, and the focal length of the lens is f.

Some typical examples are the following:

Extension tube: Focal length:	E = 10 mm f = 50 mm		E = 15 mm f = 135 mm		E = 25 mm f = 300 mm	
	Lens setting S	Object distance P	S	P	S	P
	∞	11.8 inches	∞	4.43 feet	∞	12.80 feet
	10 feet	11.1	20 feet	3.76	50 feet	10.50
	5	10.5	10	3.26	35	9.75
	2	8.8	6	2.76	25	8.90

Conversely, if we wish to focus on an object at a distance P, the second formula tells us what lens setting will be required with any given extension tube and focal length. It should be remarked that not all lenses behave well at short object distances, and when there is an actual image magnification the lens should be reversed so that light enters the rear and emerges from the front. Of course, with an SLR camera no calculations or tabular data are required, and any extension tube can be used with any lens.

Telenegative Attachments

Many companies are now making 2× and 3× telenegative attachment lenses. These are intended to be inserted between a regular lens and the camera, where they double or triple the focal length and also the F-number of the main lens. Thus, inserting a 2× telenegative attachment behind a 50 mm $f/2$ lens gives the photographer a 100 mm $f/4$ combination, or, when used behind a 35–85 mm $f/2.8$ zoom lens, it gives the user a 70–170 mm $f/5.6$ zoom for a very small additional cost.

Such a combination can actually be more useful than a normal lens of the longer focal length, because a 100 mm lens can rarely be focused closer than about 3.5 feet, while the combination focuses as close as the main lens, namely, about 1.5 feet. The barrel containing the negative attachment has just the right thickness so that when the main lens scale reads ∞, the combination is correctly focused for a very distant object. Often the automatic iris diaphragm controls are extended through the negative attachment, so that the automatic features of the camera continue to operate when the attachment is in place. Provided it is well designed, such an attachment does not degrade the image quality to a noticeable extent.

The Vivitar Company makes a 2× telenegative attachment called a "2× Macro Focusing Teleconverter" that enables the main lens to be moved forward by about 0.6 inch, so that when combined with the maximum focusing range of the main lens an overall magnification of about unity can be achieved, great for copying small objects such as postage stamps. However, as the main lens is now working at about 0.5×, considerably beyond its design magnification of about 0.12×, it is desirable to stop down the system when working at such an extreme demagnification.

Diffusion Attachments

Some photographs may be substantially improved by the use of a little diffusion, either on the camera or the enlarger. Diffusion is the scattering of a fraction of the incident light into a narrow cone surrounding the image-forming rays, thus lowering the detail contrast of the picture and softening the sharp edges of objects. Diffusion should not be confused with an out-of-focus condition, because in that case there is no sharp image at all, or worse still, the wrong things are in focus, whereas when diffusion is used most of the light still forms a sharp image that is superposed on the diffuse image.

Diffusion may be obtained by merely inserting a piece of silk stocking or other fine transparent cloth between the enlarging lens and the printing paper. However, many workers prefer to use a glass *diffusing disk* over the lens in a regular filter holder. This disk is a parallel plate of glass on which a pattern of lines or circles has been produced by a local polishing action. The polisher produces a shallow cylindrical groove on the glass that spreads out the light falling on it into a cone, as indicated in Fig. 8.11, while most of the light passes unaffected through the plate. The degree of diffusion is therefore determined both by the depth of the polished grooves and the percentage of the plate that is occupied by them.

Various other methods have been tried to produce a practical diffusing

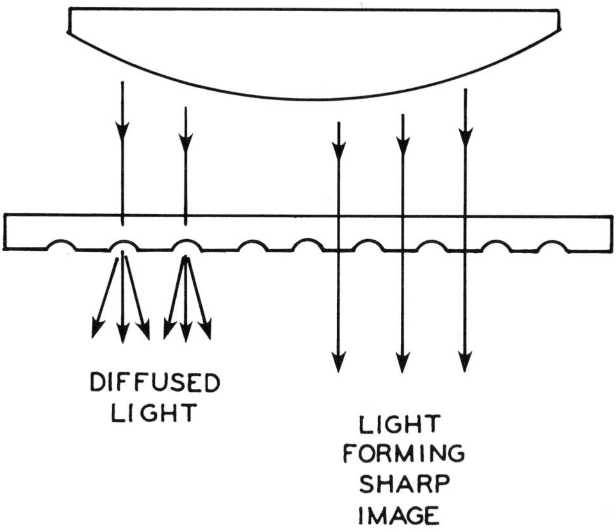

Figure 8.11. The action of a diffusing disk in enlarging.

disk, one being to fuse a pattern of glass threads between two glass plates having a different index of refraction from that of the threads.

It should be noted that there may be a significant difference between the results obtained when the same diffusing plate is used either over the camera lens or over the enlarging lens. When diffusion is used in the camera, light is scattered into the shadows, whereas in the enlarger the highlights are darkened. Since diffusion is an aesthetic question, the preferred use and degree of diffusion will depend entirely on the type of subject and on the preferences of the photographer, and no fixed rules or recommendations can be laid down. On the whole, it seems best to use diffusion on the camera, but is actually much more convenient to have a sharp original negative that may be subsequently diffused in printing to any desired extent.

A trace of diffusion is often useful in extreme enlargements to remove graininess and marks on the negative, and to produce the effect of a slight atmospheric haze in outdoor shots. Extreme diffusion is rarely used today, although it has had waves of popularity in the past.

The degree of diffusion obtained from a given diffusing disk can be reduced by using the disk for only a part of the exposure. However, this procedure is a risky one as the image through the disk may not coincide perfectly with the direct image. For example, if the plate is slightly wedge-shaped, it will displace the entire image to one side by a few tenths of a millimeter, thus causing a doubling of the image. The parallelism of well-made diffusion disks is so carefully controlled, however, that this effect is not very likely to occur. A second and more subtle trouble is due to the finite thickness of the plate itself, as is shown diagrammatically in Fig. 8.12. Here it will be seen that the insertion of a thick parallel plate of glass between the lens and easel of an enlarger throws the image slightly out of focus and produces a slight radial reduction in magnification. Thus, if the disk is used for only part of an exposure and then removed entirely, there will be a progressively increasing radial doubling of the image toward the outer parts of the field. Both these radial and lateral displacements of the image may conceivably occur together, which would lead to an unpleasant unsymmetrical type of degraded definition.

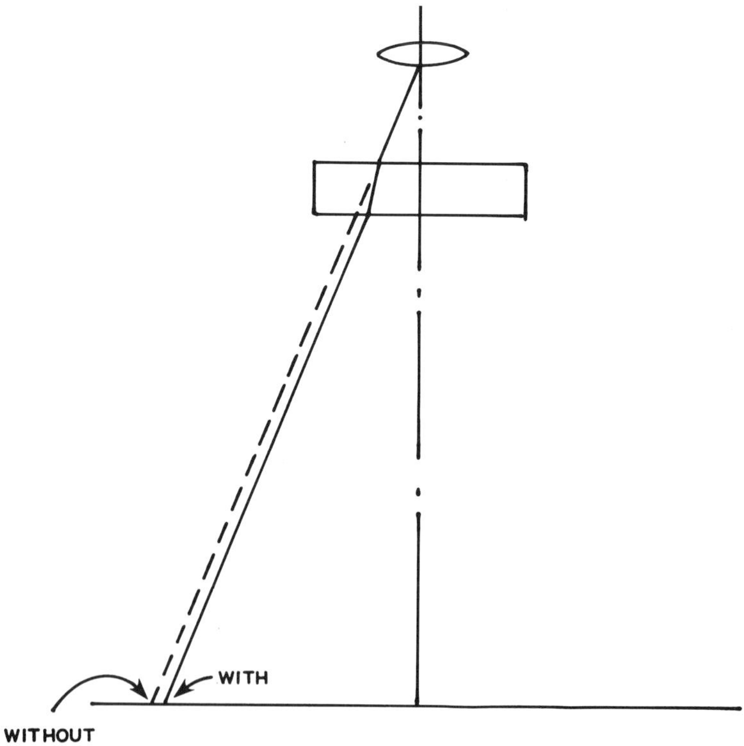

Figure 8.12. Radial displacement of an image caused by a thick filter or diffusing disk.

The kind of diffusion obtained with a diffusion disk differs from that given by a soft-focus lens, in which the softening is obtained by the deliberate introduction of large amounts of spherical aberration. In the soft-focus lens the depth of field is greatly increased by the aberration, which is not the case when a diffusion disk is used. Professional photographers often prefer the soft-focus lens for this reason, and also because the degree of diffusion can be controlled by adjusting the iris diaphragm.

Photography through a Telescope

Anything that can be seen can be photographed, including the image seen in a telescope. It is only necessary to focus the telescope on the subject, and then replace the eye by a camera focused on infinity.

Two cases now arise: the exit pupil (Ramsden disk) of the telescope may be either *larger* or *smaller* than the entrance pupil of the camera lens, and whichever of these two apertures is the smaller will become the limiting aperture of the complete system. If the camera aperture is smaller than the Ramsden disk, then the effective *F*-number of the camera will be unchanged by the addition of the telescope, and the normal exposure can be given with a small allowance for the loss of light in the telescope. On the other hand, if the Ramsden disk is smaller than the camera lens, as often happens with a relatively high-power telescope, then the Ramsden disk becomes the limiting aperture and the *F*-number of the system will be equal to the camera-lens focal length divided by the diameter of the Ramsden disk. The exposure time must be computed on the basis of this low relative aperture, plus a small allowance for the loss of light in the telescope itself.

Whenever a camera is used behind a telescope in this way, the greatest care must be taken to ensure that the Ramsden disk of the telescope is accurately located in the plane of the entrance pupil of the camera lens, both laterally and longitudinally; otherwise the entire field of view will not be recorded. Similarly, the angular field of the camera must be equal to or greater than the angular field of the telescope to ensure that the field will all be recorded on the film.

For nature and bird photography, it is helpful to mount on a tripod a pair of binoculars of the type having center-knob focusing, with a still or movie camera behind one side of the binoculars, leaving the other side free for use as a most convenient focusing viewfinder. Suppose the binocular magnifies 6 times and is used in front of a one-inch cine lens. The result would be equivalent to using a 6-inch lens on the cine camera. But the viewfinder provided with most cine cameras for use with a 6-inch lens usually has a very small apparent field and no means for accurate focusing, so that the apparently cumbersome binocular arrangement has great practical advantages.[a]

It is possible to take photographs directly through a telescope without using a lens on the camera, by racking out the eyepiece until it projects a well-focused real image upon the ground glass screen. The F-number of the system in this case is found by dividing the distance from the exit pupil of the telescope to the film by the diameter of the exit pupil itself. Thus, if a 3-inch 50-power astronomical telescope is used to project a real image at a distance of 5 inches from the exit pupil, the relative aperture of the system will be (5 inches)/(3/50 inch) = $f/83$. The exposure time must therefore be suitable for a lens of this low aperture.

The calculations and allowances listed here are needed for a rangefinder camera, but not, of course, for a single-lens reflex camera where everything is taken care of automatically.

Photography through a Microscope[b]

All that has been said about photography through a telescope applies equally well to photography through a compound microscope by placing a complete camera behind the eyepiece, or by defocusing the microscope and projecting a real image directly upon the film. The aperture relations stated for a telescope apply also to a microscope. Since the exit pupil of a microscope is generally very small, exposure times are often long, and the depth of focus at the film is often surprisingly great.

Afocal Attachment Lenses

Before the introduction of zoom lenses, several manufacturers provided low-power afocal Galilean telescopes, either direct or reversed, mounted so that they could be easily added in front of a regular camera or movie lens. These have the effect of increasing or reducing the focal length of the original lens, without affecting in any way either its F-number or back focal distance, by merely shifting the principal plane toward or away from the film (Fig. 8.13).

It might be supposed that the same attachment could be used either way to produce an increase or a reduction in the focal length at will, but the

[a]J. H. McLeod, "Telephoto Motion Pictures through Binoculars," The Camera **47**, 301 (1933); **48**, 327 (1934).
[b]R. P. Loveland, *Photomicrography*, Wiley, New York (1970).

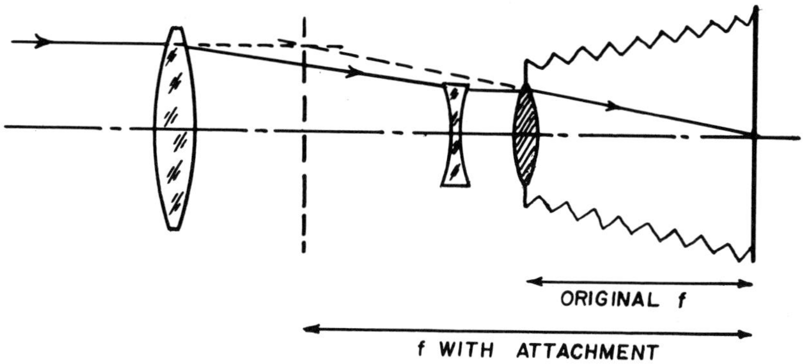

Figure 8.13. The use of an afocal (telescopic) attachment to increase the focal length of a camera lens.

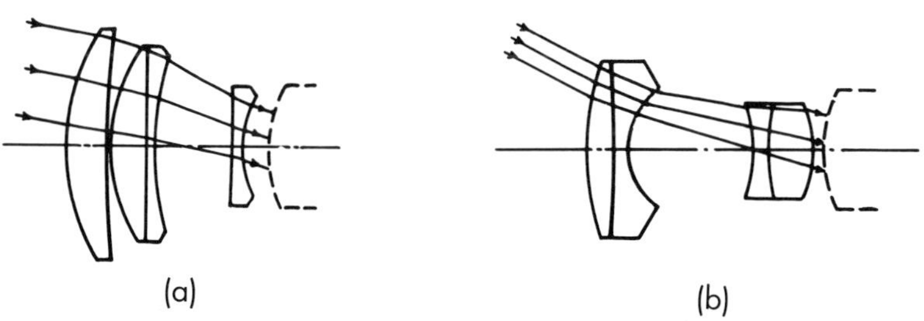

Figure 8.14. Typical afocal telescopic attachments: (a) telephoto attachment and (b) wide-angle attachment.

sizes and internal constructions of the positive and negative components must be quite different in the two cases. Typical forms of these "wide-angle" and "telephoto" attachments are shown in Fig. 8.14. It will be noticed that the oblique pencils between the two components are nearly parallel to the lens axis in the wide-angle attachment, but they are steeply inclined to the axis in the telephoto attachment; thus, the front component in the telephoto case must be relatively much larger than in the wide-angle case. The attachment must be mounted very close to the front face of the main lens to reduce vignetting.

Although afocal attachments have never been extensively used on 16mm motion-picture cameras, they have been extremely popular since World War II in the 8mm field. The most usual procedure has been to equip the camera with a three-hole turret in front of the lens (Fig. 8.15), two of

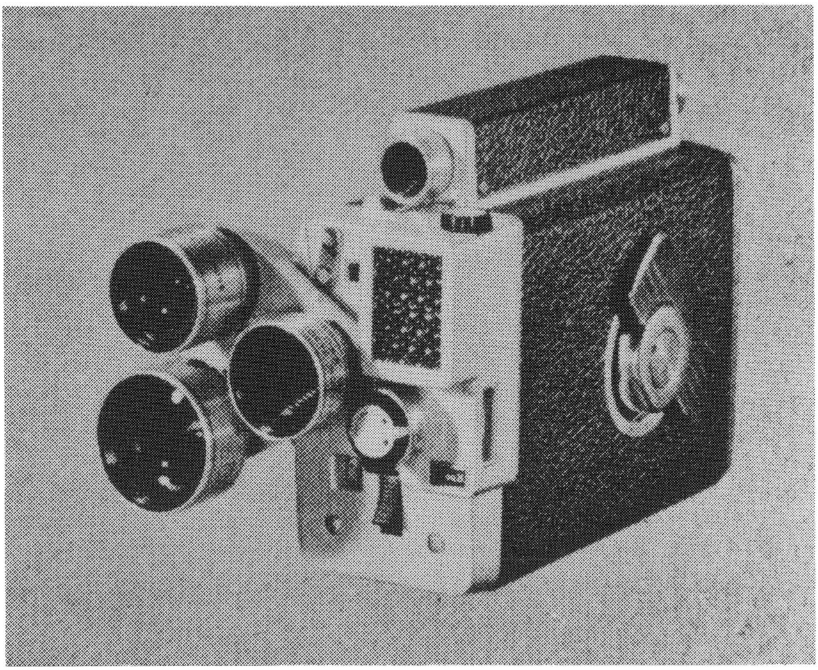

Figure 8.15. An 8mm camera equipppped with a turret carrying two afocal attachments.

the holes containing afocal attachments, one telephoto and one wide-angle, the third hole being empty. The diaphragm or wheel of stops is usually placed immediately behind the turret, in front of the main lens of the camera. The effect of such an afocal attachment on the focus scale of the camera, if there is one, and on the depth of field, has been discussed on page 176.

A Fish-Eye Attachment

An interesting example of an afocal (telescopic) supplementary lens is a fish-eye attachment. One example is made by Samigon in Japan. The axial entrance pupil diameter is about 1.2 mm and the exit pupil diameter is 8.5 mm, giving an axial magnifying power of 1/7.08. The iris diaphragm is placed close to the rear of the attachment so as to have the least possible effect on the aberration correction of the main lens, which of course must be fully open when the attachment is used.

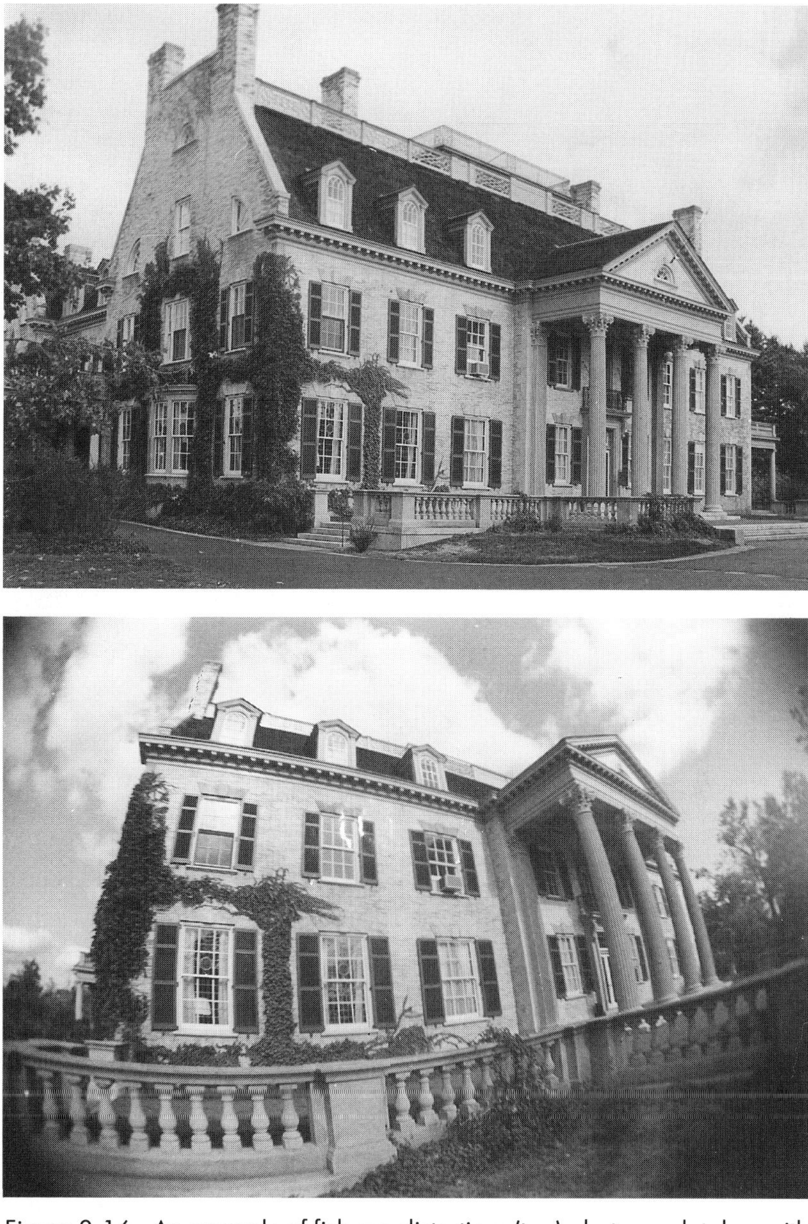

Figure 8.16. An example of fish-eye distortion: (top) photograph taken with a regular 20mm lens, and (bottom) photograph taken with a fish-eye attachment mounted on a 100mm lens.

The iris scale of the Samigon attachment has a maximum reading depending on the focal length of the main lens, indicating *f*/5.6 when a 50mm lens is used and *f*/11 if the main lens is 100 mm. The focal length of the combination is, of course, equal to 50/7.08 = 7mm for the 50mm lens, or 14 mm with the 100mm lens. In each case the diameter of the entrance pupil works out to be 1.2 mm. The circular image on the film has a diameter of 20 mm if the 50 mm main lens is used, or twice that amount with the 100mm lens, although it is difficult to determine this as the image is larger than the 24 × 36 mm frame.

Figure 8.16 shows two different pictures of a building. The upper view was taken with a regular 20mm lens, and the lower view with the fish-eye attachment mounted on a 100mm lens. For the fish-eye view, the camera was placed close to the stone balustrade, which appears curved in the picture although it was actually quite straight. For the normal lens it was necessary to move back many feet in order to get the whole building into the picture. It is noticeable how the fish-eye system has stressed the clouds. This is due to the huge barrel distortion in this type of system, which has the effect of increasing the illumination in the outer parts of the field.

Anamorphic Attachments

If an afocal telescopic attachment system of the type described in the previous section is made with cylindrical surfaces instead of the usual spherical surfaces, the axes of all the cylinders being parallel, then the attachment will magnify or diminish the image only in a direction perpendicular to the cylinder axes. This has the effect of stretching or compressing the image in that direction only, giving some degree of anamorphic distortion. No significant use has been made of this device on still cameras, but it was proposed by Zollinger as early as 1910 that a cylindrical-lens anamorphoser should be mounted in front of a motion-picture camera, to compress the image vertically and thus save film, a similar anamorphoser being used in front of the projection lens to reexpand the picture vertically and restore it to its original proportions. The first commercial application of this principle was the very successful CinemaScope system introduced by Twentieth-Century Fox in 1952, except that here the compression is horizontal so as to reduce a very wide picture to the ordinary 3:4 proportions of a standard motion-picture camera or projector. The original lens used for this purpose was the Hypergonar designed by H. Chrétien in 1929 (Fig. 8.17).

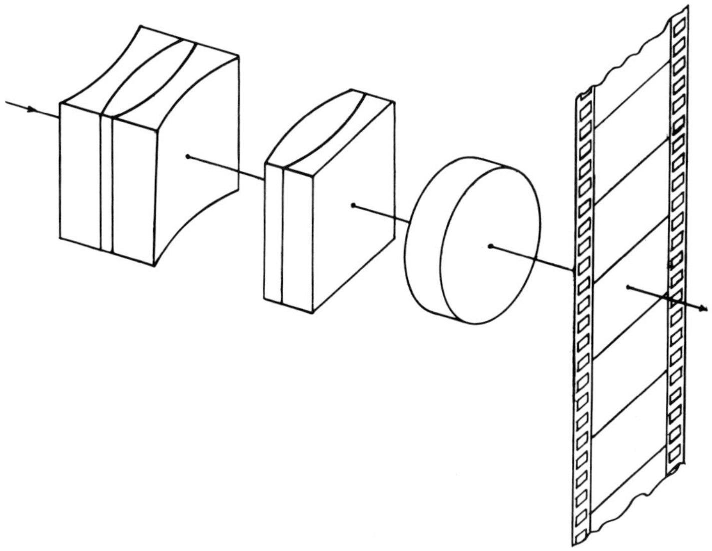

Figure 8.17. Chrétien's Hypergonar lens.

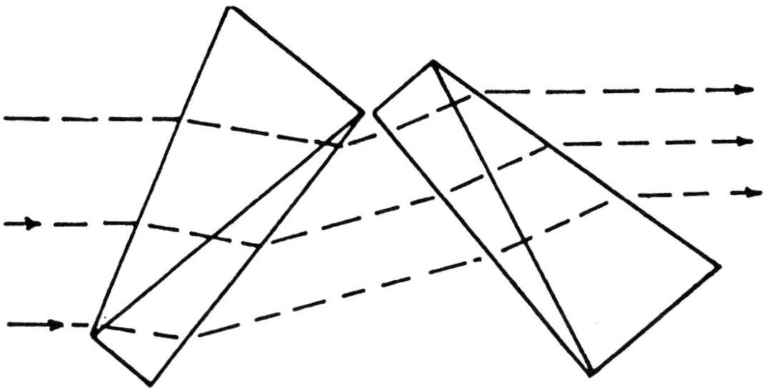

Figure 8.18. A Brewster prism anamorphoser.

Anamorphic compression of this type can also be obtained by a pair of refracting prisms arranged as shown in Fig. 8.18, in which the prisms are both tilted away from the mid-position, by equal amounts but in opposite directions. The *deviation* of the light by one prism is thus rectified by the other, but each prism contributes its share to the anamorphic compression. Prisms are preferred to cylindrical lenses where the field angle is small and the aperture is large, as in motion-picture projectors, but for the comparatively wide field covered by a camera, cylinder-lens attachments are generally preferred.

When the object, or projection screen, is at a finite distance, it is essential to focus the anamorphoser in some way so that the light rays in both the compression meridian and in the neutral meridian will come to a common focus. Since prisms cannot be focused, it is customary to use an adjustable collimator consisting of a pair of separated positive and negative lenses in front of a prism anamorphoser; the collimator must be adjusted so that truly parallel light passes through the tilted prism.

Chapter 9

Enlarging and Projection Systems

Projection Printing

Although photographic printing by contact is the simplest procedure that could be devised for the purpose, there are several good reasons why printing by optical projection is often preferable.

By optical projection we can secure any desired magnification or demagnification, and thus make finished prints of any desired size from the whole or part of a negative. The development of the miniature camera would have been impossible without enlargers to provide prints of a useful size. Moreover, enlarged prints appear much more natural than contact prints because their center of perspective lies closer to the observer's eye.

In addition to this, projection printing gives the operator easy control of the exposure by "dodging," and it enables one to correct the effects of a tilted camera by suitably orienting or tilting the easel. Projection printing has some further advantages: for instance, the negative may be suspended out of contact with other matter to avoid damage to a delicate film and to avoid the Newton's rings and dust spots that arise when a film is clamped between glass plates.

Projection printing is often used to prepare direct copies on bromide paper of letters or printed matter, where a contact print would be impossible* as

*This difficulty was overcome in the Kodak Verifax system by the device of exposing the copy material, not through the matter to be copied but through the back of the sensitized paper. This paper was of high contrast with a sharp "toe" to the characteristic curve, and the exposure time was made too short to fog the paper on the forward passage of light through it. However, after reflection from the white background paper, the light was strong enough to go past the toe and blacken the emulsion, whereas light from the letterpress was insufficient to do so. The result was a negative image that was reversed chemically to produce the desired positive reproduction of the material.

the lettering on both sides of the sheet would be simultaneously recorded. The familiar Photostat and Xerox machines operate on this principle.

Finally, the type of illumination employed in projection printing has some effect on the contrast of the print. Thus, with a condenser enlarger we can lower the contrast by inserting an opal glass plate behind the negative, but the difference will scarcely amount to more than the difference between one grade of bromide paper and the next.

In spite of these advantages, some definition is always lost when an enlarger is employed, and no projected print can approach in brilliance of detail a good contact print on glossy paper.*

Magnification in Projection Printing

The essential formulas connecting the conjugate distances with the magnification have already been given (page 33). They are as follows:

$$a = f\left(1 + \frac{1}{m}\right) ,$$

$$b = f(1 + m) ,$$

$$D = a + b = f\left(2 + m + \frac{1}{m}\right) = f\,\frac{(1 + m)^2}{m} ,$$

$$f = a\,\frac{m}{1 + m} = \frac{b}{1 + m} = \frac{mD}{(1 + m)^2} ,$$

$$m = \frac{b}{a} = {}^1\!/_2\,(k \pm \sqrt{k^2 - 4k}) - 1 , \quad \text{where } k = \frac{D}{f} .$$

In these formulas, a and b are the conjugate distances from object and image to their respective principal points in the lens (considered positive if they are real distances), f is the focal length, D is the overall object-to-image distance, and m is the magnification, again considered positive for a real image. The separation between the principal points has been neglected here.

*An excellent summary of the history of photographic enlargers has been given by E. Ostroff in Photo. Sci. & Eng. **28**, pp. 54–89 (March–April 1984).

Thus, using a 4-inch lens to obtain a linear magnification of 3×, the film-to-lens distance a is equal to $4(1 + 1/3) = 5.33$ inches, and the lens-to-paper distance b is given by $4(1 + 3) = 16$ inches. The total height of the enlarger from film to paper is therefore $D = a + b = 21.3$ inches.

For some special purposes such as the commercial printing of amateur negatives, the distance from negative to paper is fixed and the width of the roll of printing paper is also fixed. Different lenses are then provided for use with different negative sizes. The above formulas can be very conveniently applied to determine the focal lengths and positions of the various lenses needed for an apparatus of this kind.

Extension of the Range of an Enlarger by the Use of Supplementary Lenses

The range of magnifications that can be covered by a given enlarger is limited at the upper end by the maximum possible distance from film to easel, and at the lower end by the least possible distance from the lens to the easel, which in turn depends both on the height of the vertical stem and on the bellows extension available. By somewhat shortening the focal length of the lens, for example by the addition of a positive supplementary lens, both limits to the range of magnifications can be considerably increased.

Focusing an Enlarger

When an enlarger is used to give a magnification considerably different from unity, the projected picture may be focused, either by moving the whole head or by moving the lens alone. However, it must be remembered that at or near unit magnification, moving the lens does not alter the focus but only the magnification. In this case, therefore, the image can be focused only by the movement of the negative or of the paper board. This point is frequently forgotten by those using common types of enlarger, which are usually focused by moving the lens.

Autofocus Enlargers

A great many ingenious devices* have been invented to maintain enlarger focus automatically, the only operating movement being one to vary the degree of enlargement. A simple device for this purpose is indicated in

*See, for instance, R. Kingslake, *Optical System Design*, p. 62, Academic Press, New York (1983).

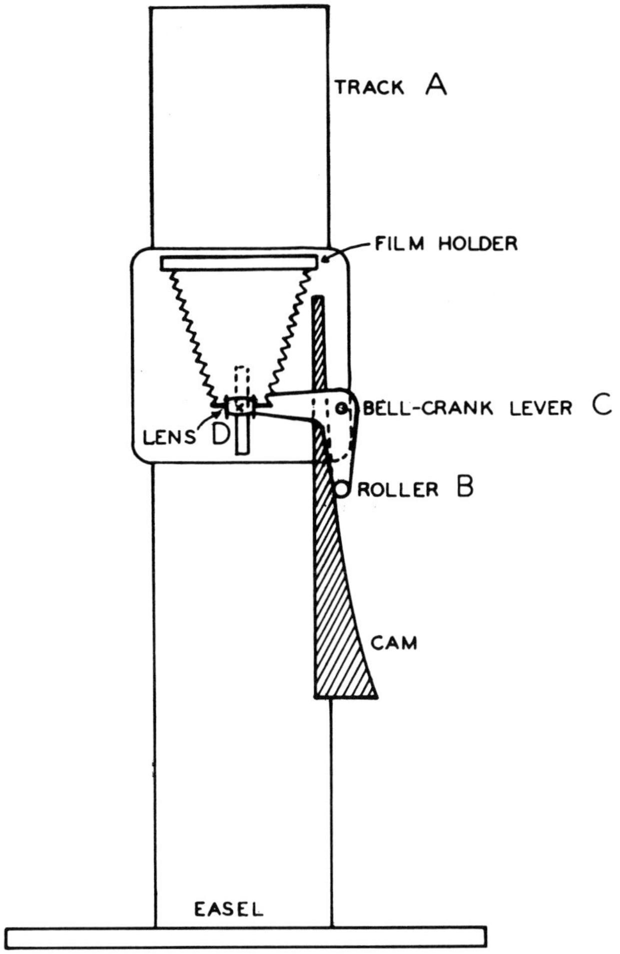

Figure 9.1. Schematic of a typical autofocus enlarger mechanism.

Fig. 9.1. The whole assembly of lamp house, negative holder, and lens slides up and down a vertical track *A*, on the side of which is a suitably shaped profile against which rides the little roller *B*. This is connected to the lens board *D* through a bell-crank lever C in such a way that as the whole assembly is lowered to reduce the magnification, the lens drops quicker than the film holder, thus serving to maintain the image in proper focus. An enlarger constructed on this principle is shown in Fig. 9.2.

It is a very difficult matter to manufacture an autofocus mechanism of this type with sufficient accuracy, especially as the focal length of the lens must be related to the other dimensions of the mechanism in a very precise

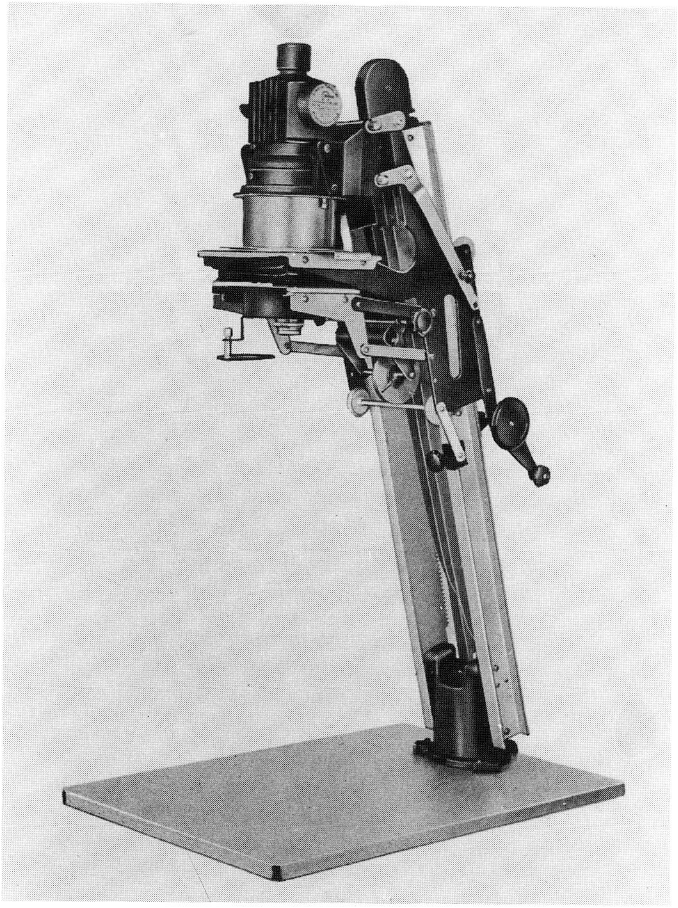

Figure 9.2. The Automega B-7 enlarger.

manner. Consequently, in spite of the many great conveniences of an autofocus arrangement, most serious workers prefer an enlarger of normal type in which the picture size and focus are determined by trial. A satisfactory compromise is one combining a rough autofocus mechanism for use in selecting the right picture size, and a focusing screw to move either the lens alone or the entire enlarger head for the final adjustment of the focus.

Enlarger Illumination

Thanks to the high speed of modern bromide papers, it is not necessary that the illumination of the negative be very bright, but it must be uniform.

Consequently, as indicated in Fig. 9.3, every enlarger contains either (a) a plate of diffusing material such as opal glass immediately behind the negative, (b) a condenser, (c) a reflective diffusing surface lit by indirect light and situated a few inches behind the negative, (d) an ellipsoidal

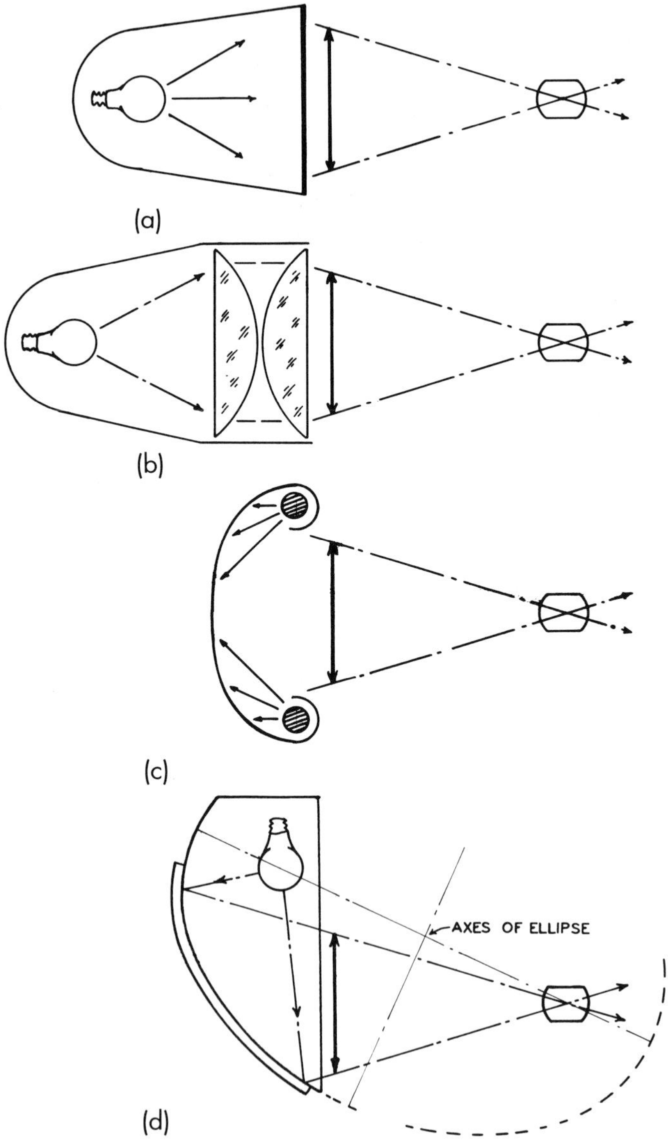

(a) (b) (c) (d)

Figure 9.3. Illuminating systems for enlargers: (a) diffuse transmission, (b) specular transmission, (c) diffuse reflection, and (d) specular reflection.

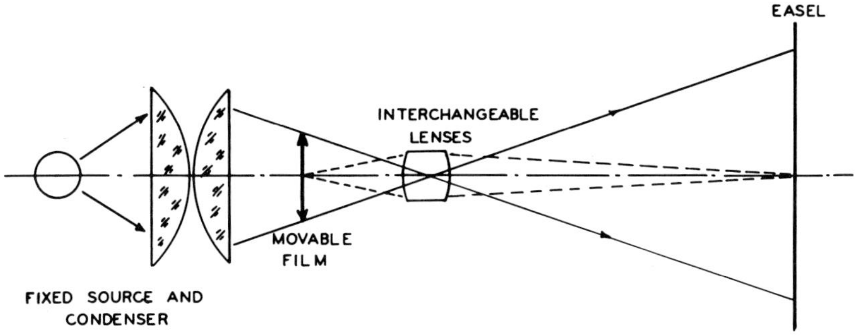

Figure 9.4. A fixed condenser enlarger for use with interchangeable lenses.

mirror with the lamp at one focus and the lens at the other, or (e) some combination of these arrangements. The directed light given by a condenser or elliptic mirror is called *specular* illumination, in contrast to the *diffuse* illumination given by the other devices.

With diffuse illumination the exposure time is likely to be long, scratches and marks on the negative tend to disappear from view, and the contrast of the negative is somewhat lowered. Extended areas of uniform tone are generally more uniform with a diffuse source, but it is common to find a "hot spot," or region of excessive brightness, in the middle of the picture if the illuminant consists of a sheet of opal glass lit from behind by a single lamp. This can be avoided by the use of several lamps or fluorescent tubes behind the opal glass. The new "cold light" enlargers are generally very satisfactory in this regard.

With condenser illumination, on the other hand, the brilliance is much increased, no hot spot is formed, and the contrast of the picture is enhanced. In practice it is essential to add some diffusion to a condenser enlarger to eliminate scratches and marks on the film, which tend to show up strongly when illuminated by a purely specular beam. This may be done either by means of a plate of finely ground glass situated between the lamp and the negative, or by the use of a lamp having a large diffusing bulb. However, if the sharpest possible definition is desired, a purely specular beam with a well-corrected achromatic condenser should be employed.

Ideally, the purpose of the condenser is to image the source into the enlarging lens. With a small source, it is therefore necessary to adjust the longitudinal position of the source to maintain uniform illumination with each change in magnification. An alternative for doing this is to have the source, condenser, and lens in a fixed relationship, and to move only the negative and the easel. In this way interchangeable lenses of different focal lengths can be used with various negative sizes (Fig. 9.4). Fortunately, with the relatively large diffusing lamps now in general use, these precautions have become unnecessary, and any film size or focal length of enlarging lens within reasonable limits can be used on a given enlarger without adjusting the lamp relative to the condenser.

The Effect of Aberrations in the Enlarging Lens

The enlarging lens performs a very similar function to that of a camera lens, and it is not uncommon to find the same lens used for both purposes. Indeed, many home-made enlargers have been built embodying the entire camera, including film holder, bellows, and lens.

It is thus worthwhile to examine the necessary properties of a lens that is to be used on an enlarger and to compare these properties with the desirable attributes of a good camera lens.

Curvature of field—If we were to reverse the direction of the light through a camera, with an inferior lens, we should find that the image of the flat film plane would be projected out into the object space as a curved surface. In photography, therefore, any object that happens to be lying on that surface will be sharply imaged on the film, and we shall not be conscious that this surface of sharp definition is actually curved. However, in enlarging, we must form an image of a flat object plane accurately upon a flat image plane, and then any field curvature will become immediately apparent. Flatness of field is thus more important in an enlarging lens than in a camera lens.

Conjugate distances—Camera lenses are generally used for object distances varying from about 3.5 feet to infinity. Enlargements go from 1:1 to perhaps 10:1, which for a 4-inch lens implies a long-conjugate distance varying between 8 inches and 44 inches. Thus, the complete range of object distances for the enlarger lies within the closest distances normally used for the camera. High-aperture lenses of a decidedly unsymmetrical construction suffer badly from field curvature and other aberrations when the magnification is less than about 10×. Enlarger lenses should therefore always be of a more or less symmetrical type of construction.

Aperture—Enlargers are generally used at apertures between $f/4.5$ and $f/16$. Camera lenses often go as high as $f/2$ or $f/1.5$. At these high apertures, the lens construction is likely to be decidedly unsymmetrical, and the corrections of the lens (especially field curvature) will fall off badly at magnifications below about 10:1. Hence, high-aperture camera lenses should never be used for enlarging.

Chromatic aberrations—The lateral color of an enlarging lens must be extremely well corrected, as the lens may be used to make color-separation negatives from a Kodachrome original. The longitudinal color must also be corrected in such a way that the position of the best visual focus (at about 555 nm) corresponds exactly with the position of the best focus for color-blind bromide paper (say at 400 nm). This requirement alone will suffice to eliminate some very satisfactory camera lenses from use on an enlarger.

Focusing procedure—Camera lenses are often designed so that they can be focused by altering the front airspace. This feature is never needed on an enlarger.

In spite of these limitations, it is nevertheless true that many good camera lenses make excellent enlarger lenses. The above factors are mentioned here as a warning, for it by no means follows that a good camera lens will always constitute a good enlarging lens.

It may be remarked in passing, although it should be obvious, that if a camera lens is used for enlarging, the front, or outside, of the camera lens must face the easel, and if reducing a large negative to make a small print, the front of the lens should again face the long conjugate.

Variation of Exposure Time with Degree of Enlargement

It was shown on page 108 that the illumination in the image formed by a lens is given by the universal formula

$$E = t\pi B \sin^2\theta' = \frac{t\pi B}{4(F\text{-number})^2(1+m)^2} .$$

Hence, for a lens of constant F-number, the required time of exposure will be directly proportional to $(1 + m)^2$, where m is the magnification ratio. Thus, in going from a 2× enlargement to a 4× enlargement, the exposure time must be increased in the ratio $(1 + 4)^2/(1 + 2)^2 = 25/9 = 2.8$ times. However, this is actually a somewhat academic argument because it is usual to determine the exposure time and the necessary degree of dodging by several trials, using various grades of paper or various types of surface, until the desired visual or artistic effect is obtained. The effects of changing aperture and magnification on the print density and exposure time may be readily computed with the aid of the circular slide rules such as those given in the Kodak Master Darkroom Dataguide.

Use of Diffusion Disks in Enlarging

In many cases, especially at high enlargement ratios , it is necessary to do something to soften the detail in the print, particularly if there are tiny marks and scratches on the film or if the negative is very grainy. This may be done by the introduction of a very slight degree of defocus, but this operation requires great care and cannot be carried far. A better way is to use controlled diffusion by a diffusion disk or by the insertion of a piece of silk or muslin between the lens and the paper board. This subject is discussed fully on page 182.

Anamorphic Enlarging Attachments

Another attachment that is sometimes of value in enlarging is a suitable anamorphoser to produce a somewhat increased magnification in one direction only. These devices are generally low-power Galilean telescopes made with cylindrical lenses. Such an attachment must be used with care or the result will be disappointing. First, the cylinder axes of the two components must be *exactly* parallel, for even a fraction of a degree of rotation of one cylinder relative to the other will ruin the definition completely. The procedure is, therefore, to set up the enlarger to give approximately the required image size. The anamorphoser is added in front of the enlarging lens, and the enlarger is refocused if necessary to form sharp images of lines lying perpendicular to the axes of the cylindrical lenses. The anamorphoser itself is then focused to give sharp images of lines lying parallel to the cylinder axes. The whole image should now be sharp all over, but it may be necessary to repeat these focusing adjustments for the best results. It is advisable to stop down the enlarger lens when such an attachment is used, to get the sharpest definition.

Use of a Tilted Easel

For many purposes, such as the restoration of converging parallel lines caused by a tilted camera, it is desirable to tilt the easel of an enlarger to produce a greater image magnification at one end of the picture than at the other end. Alternatively, the easel may be left horizontal and the entire enlarger tilted by the desired angle. At small apertures, the depth of field will generally be great enough to give acceptably sharp definition over the entire field, but when larger apertures must be used, it is necessary to tilt the lens also, so that the film plane and the easel plane meet on the principal plane of the lens (see page 36). Some enlargers are equipped with means for conveniently tilting the film or the lens or both for this purpose (Fig. 9.5). It is worth noting that if the enlarger or easel is tilted very much, the exposure will be noticeably greater at the end nearest to the lens, and some dodging may be required to equalize it.

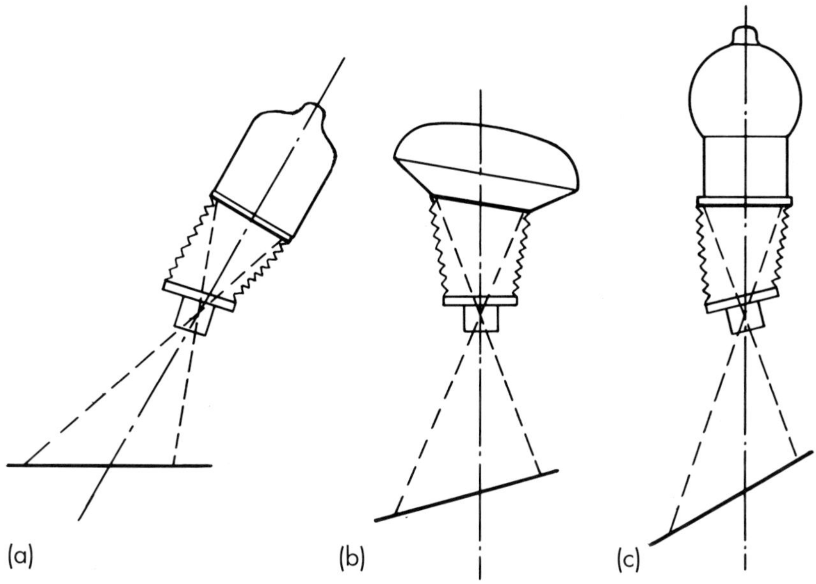

Figure 9.5. Tilting devices on enlargers for rectifying converging parallels: (a) fixed easel, tilting negative and lens, (b) fixed lens, tilting negative and easel, and (c) fixed negative, tilting lens and easel.

It is a straightforward matter to determine in advance the slope at which the easel and lens board must be set relative to the axis of the enlarger. Figure 9.6 shows a possible situation in which a tilted camera is used to photograph a tall building. Parallel vertical lines in the building appear to converge to a vanishing point V on the negative, V being the conjugate image of an infinitely distant object point located high above the building, for which the magnification is zero. The point V is, therefore, the intersection of the film plane and a line drawn through the lens parallel to the building.

To rectify these converging lines, we imagine a plane sheet of printing paper placed much closer to the camera than the original building, such as at P or P' in Fig. 9.6, and we project the tilted negative back through the lens onto this sheet of paper.

As most enlargers have a vertical axis, we rotate the diagram counterclockwise until the camera axis coincides with the enlarger axis (Fig. 9.7). The vanishing point falls at V, the camera lens is replaced by the enlarger lens, and the easel replaces the plane section P. The line joining the

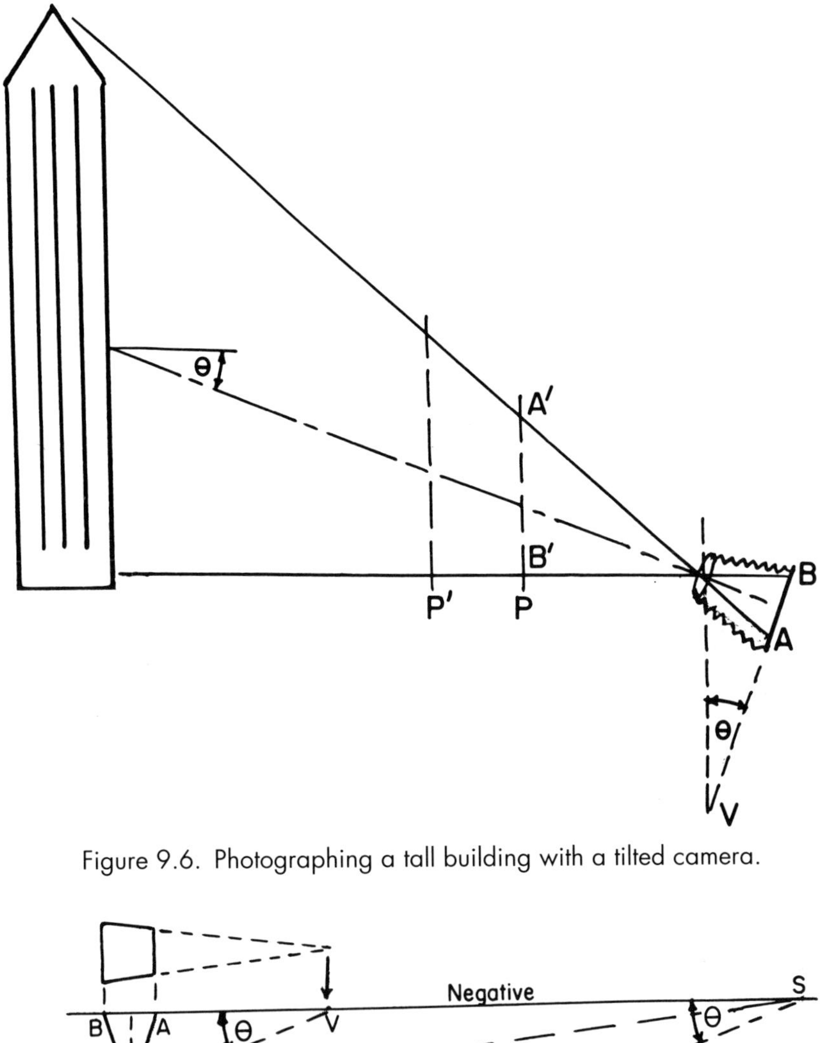

Figure 9.6. Photographing a tall building with a tilted camera.

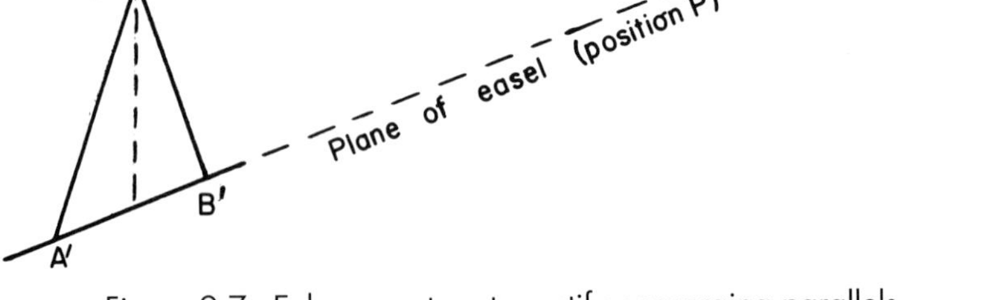

Figure 9.7. Enlarger set up to rectify converging parallels.

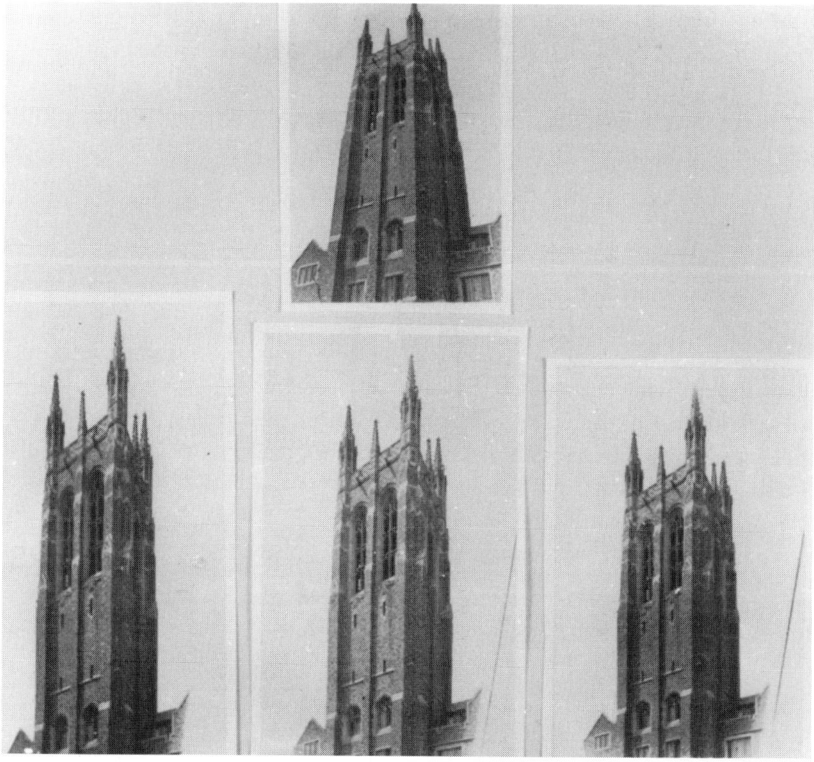

Figure 9.8. An example of anamorphic distortion caused by incorrect rectification. (Above) Original picture made with a tilted camera having a 103mm lens. (Left) Incorrect rectification using a 100mm enlarger lens. Easel tilt 47°, lens tilt 17°. (Middle) Correctly rectified enlargement made with a 73mm lens. Easel tilt 38°, lens tilt 13°. (Right) Incorrect rectification using a 56mm lens. Easel tilt 31°, lens tilt 10°.

vanishing point to the camera lens was at an angle θ from the film plane, and this angle must be reproduced in the enlarger. Hence, the distance from the negative to the enlarger lens must be equal to the focal length of the original camera lens. Thus, if the camera lens had a focal length of 4 inches, and an enlarger magnification of 3× is desired, the focal length of the enlarger lens must be 3 inches. This requirement leads to a strange set of focal lengths if a wide range of magnifications is desired, and unless a zoom lens is used it is unlikely that these focal lengths will be available.

The image magnification is, of course, greater at the lower end of the easel, at A', and smaller at the upper end at B'. By "magnification" in the conjugate distance formula is meant the magnification close to the lens axis. It is commonly measured for a transverse distance on the negative, at right angles to the vertical dimensions of the original object.

It should be noted that for good definition throughout the print, it may be necessary to tilt the enlarging lens also, to fulfill the Scheimpflug condition (see page 36). The lens tilt will ordinarily be small, about equal to the easel tilt divided by the magnification, but the improvement in definition at the extreme ends of the picture will be startling, and such a lens tilt is highly desirable if it can be achieved.

Failure to observe any of these stated conditions will lead to some degree of anamorphic distortion in the final print. As an example, Fig. 9.8 shows a print from the original negative of a high tower, taken with a 4-inch camera lens at a tilt of about 38°. The three rectified images below were made using three different enlarger lenses, of 4 inches focal length at the left, 3 inches in the middle, and 2 inches at the right. The transverse

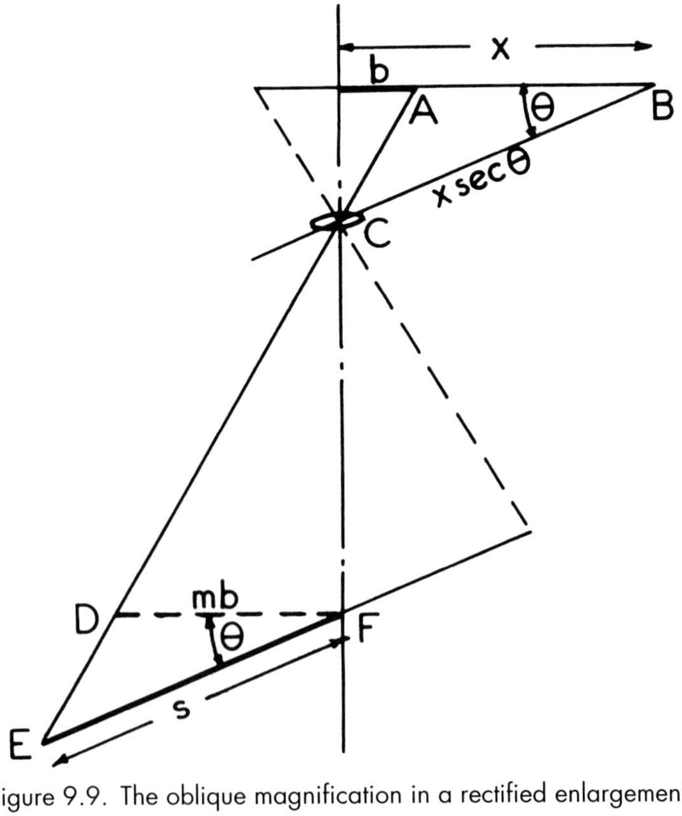

Figure 9.9. The oblique magnification in a rectified enlargement.

magnification in the middle of the negative is the same in all three cases. Only the middle picture gives a correct reproduction of the original tower.

It is interesting to note that in Fig. 9.9 the triangle *DEF* at the easel and triangle *ACB* at the negative are similar, and consequently the length of the image on the easel is given by

$$\frac{s}{m\,b} = \frac{x \sec \theta}{x - b}$$

As we are assuming that *m*, *b*, and *x* are fixed quantities, it is clear that *s* is proportional to the sec θ. Thus, the anamorphic stretching or compression resulting from the use of an incorrect enlarging lens will be in the ratio $\cos \theta_0 / \cos \theta$, where θ_0 is the slope of the camera lens and θ is the slope of the line *VL* in the enlarger.

Projection Systems

Two-Lens Projection System

The ordinary condenser enlarger system, illustrated in Fig. 9.4, is a familiar example of the well-known two-lens projection system that has been used for a great many years in "magic lanterns" and in the more modern slide projectors and home motion picture projectors.

Essentially this system provides a broad, uniformly bright area in which the slide or film can be placed, even though a highly nonuniform source of light such as a projection filament lamp is used. As has been mentioned already, the condenser is arranged to project an image of the source of light into the aperture of the projection lens, which in turn images the slide upon the screen. The whole system is shown diagrammatically in Fig. 9.10. Cones of rays originating at the slide meet again on the screen,

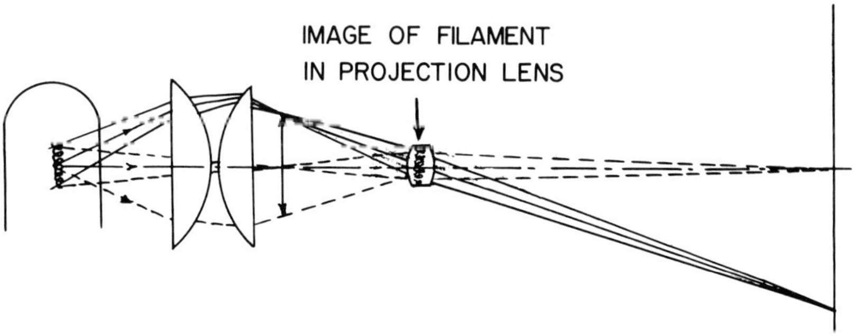

Figure 9.10. The two-lens projector system.

and it can also be seen that rays originating at the lamp, such as those marked with arrows, meet again at the projection lens and form an image of the filament there.

Since the source is imaged into the projection lens by the condenser, an observer standing at the screen and looking back toward the projector will see the lamp filament apparently situated in the projection lens aperture, and if the condenser has been well designed the filament image will fill the projection lens with light (Fig. 9.11). This now acts as a secondary source, having an intrinsic brightness equal to the brightness B of the filament itself multiplied by the transmittance of the optical system,

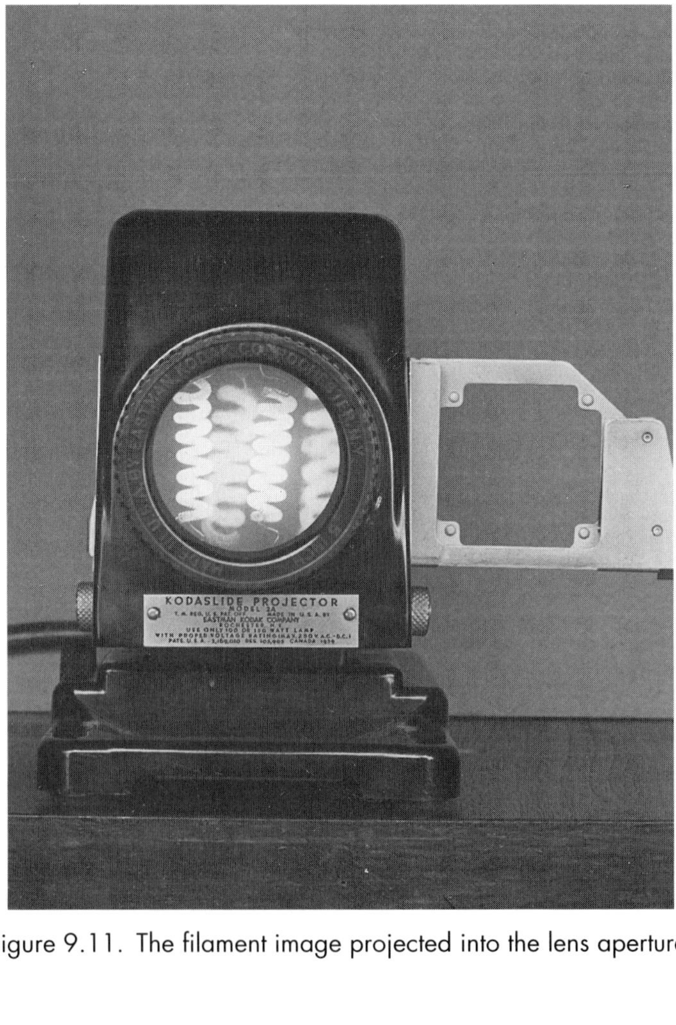

Figure 9.11. The filament image projected into the lens aperture.

as shown on page 106. Thus, if the image magnification is m and the F-number of the projection lens is N, the illumination falling on the center of the screen is given by the usual formula

$$E = t \pi B \sin^2 \theta' = \frac{t \pi B}{4 N^2 (1 + m)^2} .$$

To illustrate the use of this formula, it was found that a certain slide projector equipped with a 1000-watt lamp and an $f/3.5$ projection lens gave an illumination of 170 foot-candles when the image magnification was 34× (with no slide in the gate). In the above equation, we have, therefore,

$$t = 0.8 \text{ (say)}$$

$$E = 170 \text{ foot-candles}$$

$$N = 3.5$$

$$m = 34$$

$$\pi = 3.14$$

where $$B = \frac{4 E N^2 (1 + m)^2}{t \pi}$$

$$= 4.08 \text{ million candles per sq. ft.}$$

$$= 44 \text{ candles per sq. mm.}$$

This is a measure of the effective brightness of the 1000-watt lamp when used under these particular conditions of service. The same formula may, of course, be used to compute the expected screen illumination for any other projection system, provided the effective brightness (luminance) of the source is known.

The efficiency of a projector will, however, fall short of this expected illumination level if the condenser is not strong enough to fill the projection lens completely with light. This can be determined very easily by lighting the lamp dimly, for example, by means of a variable transformer, and looking into the projection lens from the front. If the filament image is much smaller than the lens aperture, it is a clear indication that the condenser is poorly designed and that the screen illumination is not as great as it could be.

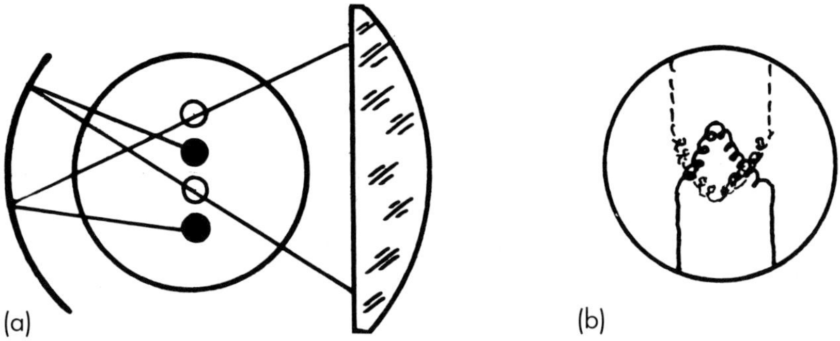

Figure 9.12. The use of a concave spherical mirror behind a projection lamp. (a) The rings are the filaments and the dots their images. (b) The appearance of the filament and its image in the projection lens.

This situation can be improved by the use of a concave spherical mirror behind the lamp, mounted so that the filament is at the center of the curvature of the mirror (Fig. 9.12). This will form an image of the filament in its own plane (clearly visible in Fig. 9.11), and effectively fill up the gaps in the filament itself, provided the mirror is carefully located in its correct position. An incorrectly mounted mirror is quite useless.

Formula (2.6) on page 33 provides the explanation for the well-known phenomenon that the projected image becomes dimmer if a 5-inch *f*/3.5 projection lens is replaced, say, by a 4-inch *f*/3.5 lens. The aperture diameter of the former lens was 5/3.5 = 1.42 inches, whereas in the new lens it is only 1.14 inches. As the lens area is less, the illumination in the middle of the screen will be reduced in the same proportion, namely, $(1.14/1.42)^2$, or to 64% of its former value. Looked at in another way, we notice that the size of the projected image is greater with the 4-inch lens than with the 5-inch, and since the total flux picked up by any *f*/3.5 lens will be the same, it is spread over a larger picture in the ratio of $5^2/4^2$; hence, we should expect the new illumination to be the proportion of 16/25 = 0.64, as computed previously.

Total Flux on Screen

By dividing the screen image into a number of squares of equal area and measuring the illumination at the center of each area, we can determine approximately the total number of lumens falling on the screen. Suppose the projected image is 4 × 6 feet and we divide it up into 24 areas of 1 square foot each; then a direct addition of the illumination in foot-candles measured at the centers of the 24 areas will be the total flux, or *screen lumens*. The average illumination that this projector may be expected to yield when forming an image of any other size is then simply equal to the total number of lumens divided by the area of the projected image in square feet.

Types of Condensers

The function of the condenser in a projector is to project an image of the light source into the projection lens aperture, and preferably to fill the lens aperture with light. The magnification at which the condenser operates is therefore equal to the ratio of the projection lens aperture to the diameter of the lamp filament. The aperture of the condenser must be somewhat greater than the diagonal of the slide or other transparency to be projected. Thus, if the source is large and the projection lens small, the condenser magnification will not be much different from unity, and if in addition the angular field of the projection is small, a relatively simple condenser will suffice. Some 8mm cine projectors use only a single lens for the condenser, although the better projectors generally use two lenses (Fig. 9.13).

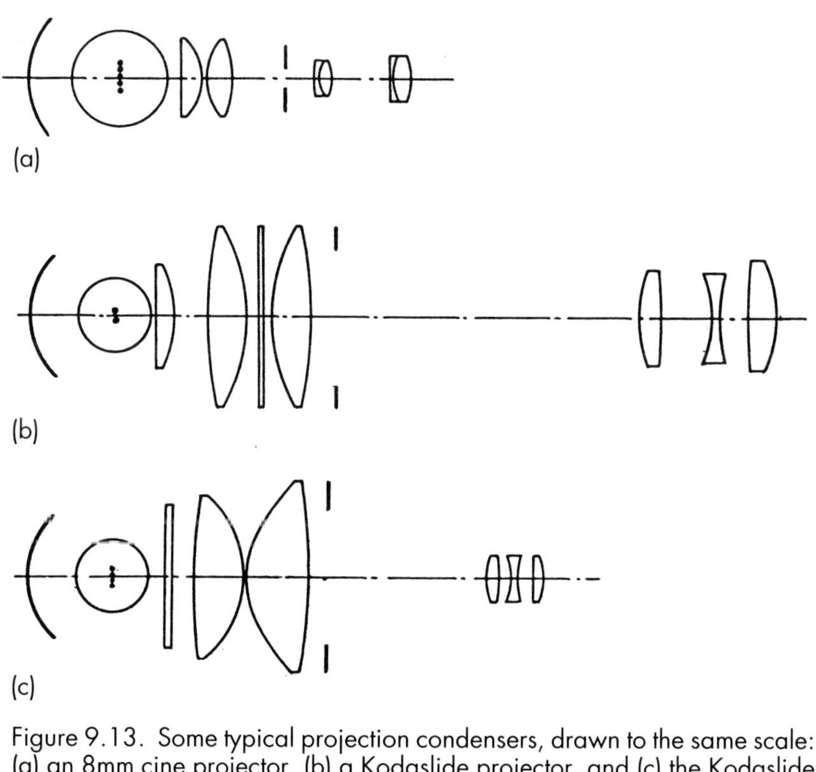

Figure 9.13. Some typical projection condensers, drawn to the same scale: (a) an 8mm cine projector, (b) a Kodaslide projector, and (c) the Kodaslide table viewer.

If a stronger condenser is required, a single lens may be divided into two or more elements, and these are generally chosen so that each surface produces about the same angular deviation of the extreme marginal ray. Occasionally, if the condenser must be very strong and also of large diameter, an aspheric surface may be introduced, a typical condenser of that type being shown in Fig. 9.13.

Lamps with Internal Reflectors

Since about 1957 projection lamps have undergone a dramatic series of changes. The older line-voltage lamps of 300, 500, 750, and 1000 watts in a tall tubular bulb [see Fig. 9.14(a)] determined the exterior form of slide projectors for many years. However, with the introduction of the newer types of short, broad lamps that can be operated in a horizontal position

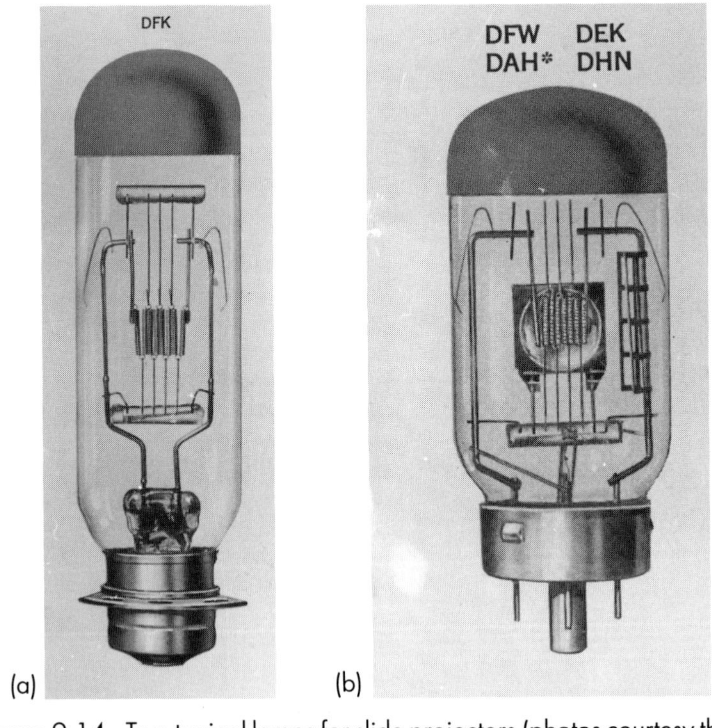

Figure 9.14. Two typical lamps for slide projectors (photos courtesy the GTE Sylvania Company): (a) An older 1000-watt upright lamp, and (b) a later 500-watt lamp that can be operated horizontally.

[Fig. 9.14(b)] containing an internal reflector, the whole design of slide projectors has changed, resulting in the modern low-format models that are now universal.

A substantial gain in projector light output has been obtained by merely substituting a low-voltage filament for the usual line-voltage filament. The low-voltage filament is made of much thicker wire, it is coiled into a more compact space, and a greater fraction of the coil is actually used in projection. Because of the increased wire thickness, it is possible to operate the lamp at a higher temperature without serious loss by evaporation, because the ratio of surface to volume is inversely proportional to the thickness of the wire. These factors taken together can easily produce a 50% increase in the screen illumination with no increase in power consumption.

Recently, a number of lamps have been developed for 8mm projectors containing an elliptical mirror of metal or aluminized glass built into the envelope (Fig. 9.15). This forms a somewhat blurred filament image close to the film gate and renders a separate glass condenser unnecessary. The

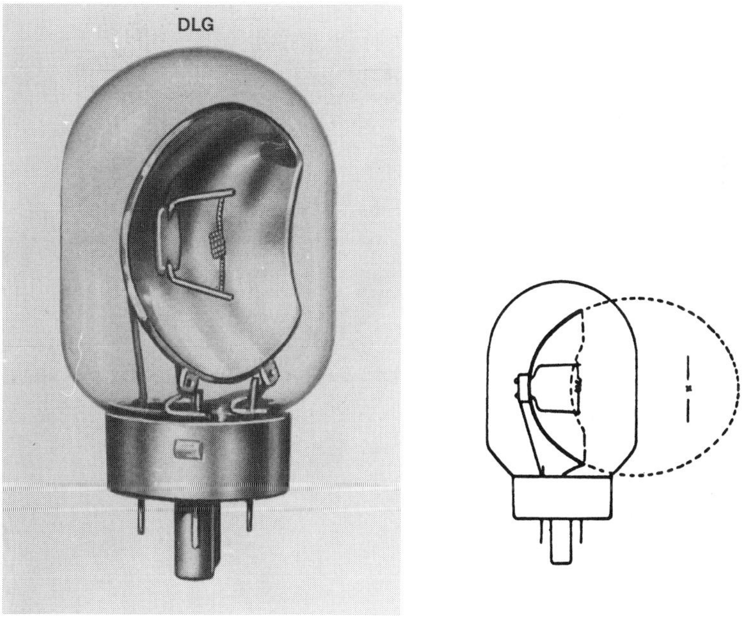

Figure 9.15. A lamp equipped with an internal elliptical reflector, and an 8mm projector gate. (Photo courtesy the GTE Sylvania Company).

mirror is large enough to fill the aperture of an $f/1.2$ projection lens. By means of these improvements, it is possible to obtain as bright a screen image with a 150-watt lamp as with the older type of 500-watt tubular lamp and condenser.

Tungsten Halogen Lamps

It was found several years ago that the introduction of iodine vapor into the lamp envelope has the property of greatly extending the life of a tungsten filament even when operated at a higher temperature. The logic of this is that the tungsten vapor that is evaporated from the hot filament combines with the iodine instead of being deposited on the inside of the envelope. The tungsten iodide vapor eventually comes into contact with the filament, where it is decomposed, and the tungsten is deposited back onto the filament and the iodine is released to repeat the cycle over and over again. Tungsten iodine lamps generally come in small envelopes of quartz or Vicor glass that are run almost red hot; the net result is a greatly increased brightness with a considerable extension of the lamp life.

The small size of the filament in these tungsten halogen lamps presents problems in condenser design. For 8mm projectors, the lamp is mounted inside an elliptical mirror that images the filament directly onto the film, as in the Sylvania Tru-beam lamps type DNE, ENX, etc. For use in a slide projector, the small halogen lamp is mounted in an elliptical mirror with the lamp filament lying along the axis of the system (Fig. 9.16). Light from the two ends of the filament then just fills the condenser lens, which is close to the slide. The condenser, in turn, forms an image of the aperture of the elliptical mirror into the projection lens. A projector of this kind is not easy to design, and some experimentation is generally needed to secure acceptably uniform illumination over the projected image of the slide. However, the efficiency of halogen lamps is high enough so that most slide projectors being designed today employ that kind of luminous source.

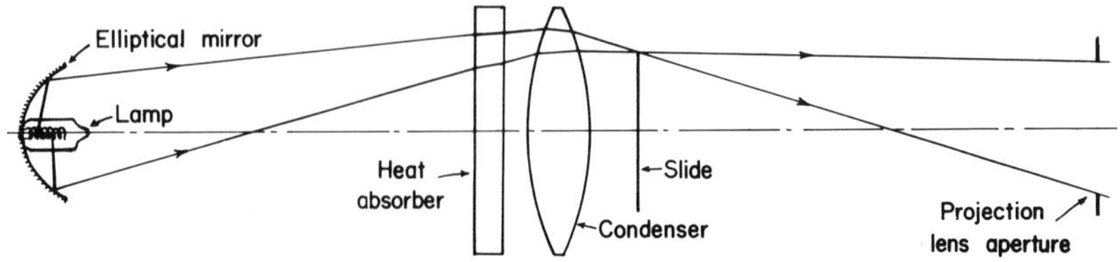

Figure 9.16. The use of a tungsten iodine lamp in a slide projector.

Distortions Caused by Oblique Projection

In most older motion-picture theaters the projection booth is situated at a considerable height above the screen, and consequently the whole projector must be tilted downward through an angle that sometimes reaches as much as 30°. If the screen is vertical, as is generally the case, three undesirable consequences arise as a result of tilting the projector, which must now be considered.

(a) The top and bottom of the picture will be slightly out of focus. This effect is small for long projection distances, and it is generally ignored.

(b) The picture is distorted anamorphically in the sense that its vertical dimension is too great. The reason for this is illustrated in Fig. 9.17, where the projected image is shown at AB and the screen at $A'B'$. The ratio of $A'B'$ to the true image size AB is approximately equal to (sec $\theta - 1$), which thus represents a measure of the relative vertical distortion. A few typical values may be of interest:

Slope θ	**Relative increase in vertical dimensions**
0°	0
5°	0.4%
10°	1.5%
15°	3.5%
20°	6.4%
25°	10.3%
30°	15.5%

It is normally considered that about 5% of this type of distortion is all that can be tolerated; hence, no theater should be constructed in which the downward slope of the projector exceeds about 18°. The black frame surrounding the screen must be made high enough to take this effect into account.

It has even been suggested that the present trend toward slim figures is an indirect result of the high projection angles of the early motion picture theaters, which made everybody on the screen look 5 or 10% taller without affecting their lateral dimensions.

(c) The width of the image of the rectangular film gate will be narrower across the top than across the bottom of the picture. This phenomenon in known as the *keystone effect*, and it is the inverse of the process of rectifying converging parallels that has been described on page 202. Referring again to Fig. 9.17, the two sides of the image converge and meet at a point P situated above the screen

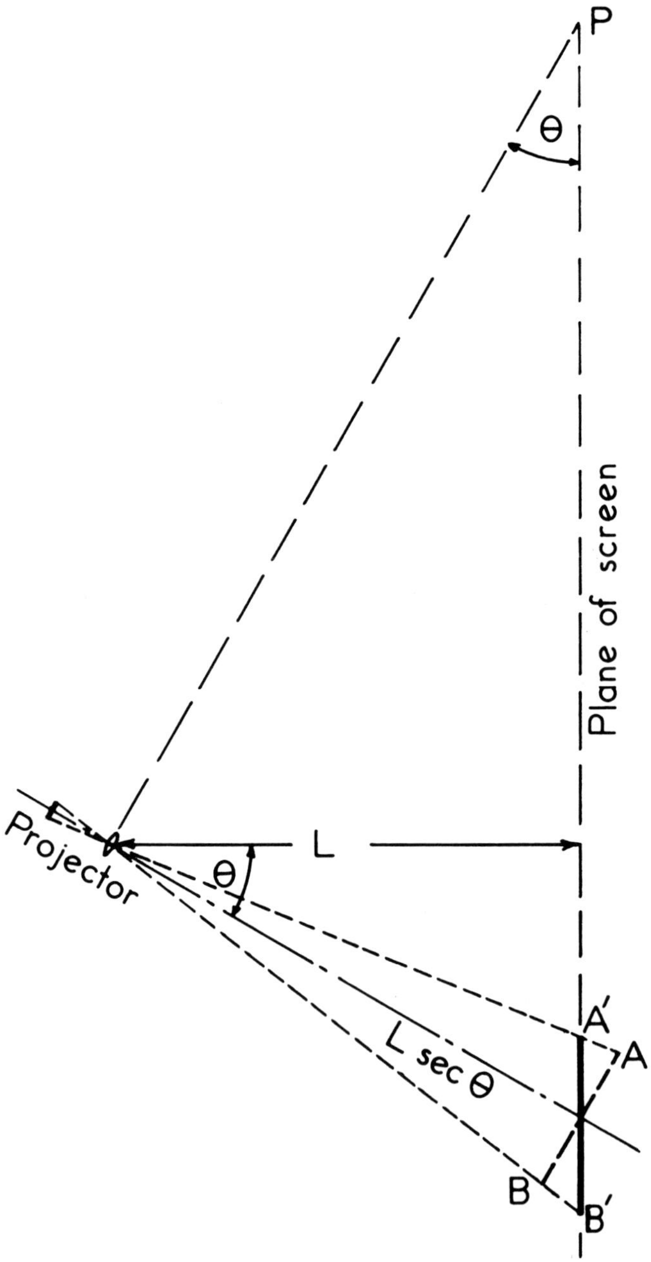

Figure 9.17. The distortions resulting from the use of a tilted projector.

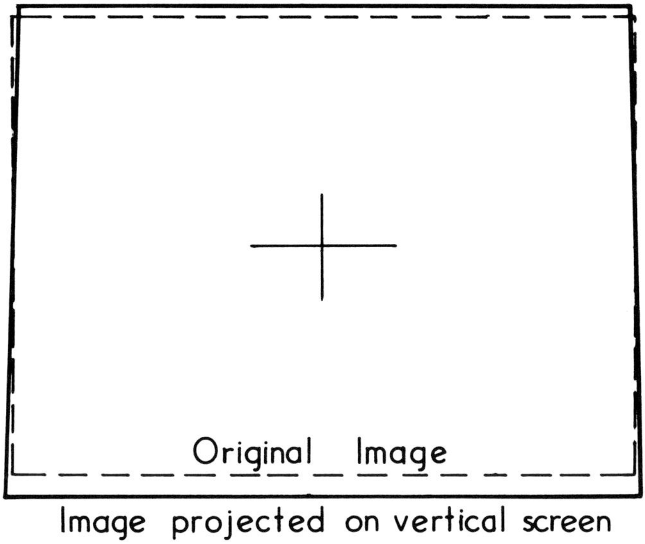

Figure 9.18. The vertical distortion and keystone effect for the cited example.

where the plane of the lens intersects the plane of the screen itself. The point P is conjugate to an infinitely distant point on the film.

If the horizontal length of the theater is L, the oblique projection distance is $L \sec\theta$, and the height of the vanishing-point P is $L/\sin\theta\cos\theta$. The magnitude of the keystone effect can thus be computed for any particular case.

Suppose, for example, a theater is 150 ft long, and the 15.25 × 20.95 mm film gate is being projected with a 6-inch lens at a 20° inclination. The oblique projection distance to the middle of the screen will be 150 sec 20° = 159.6 ft, the magnification will be 319, and the inclined rectangular image will measure 16.0 × 21.9 ft. However, when this image falls upon a vertical screen, its mean horizontal dimension will remain 21.9 ft, but its vertical dimension will be increased 6.4% to 17 ft. The height of the vanishing-point P is 159.6/sin 20° = 467 ft, and the length of the upper edge of the picture will become less in the proportion (467 – 8.5)/467 = 0.9817, and the lower edge will become greater by the ratio (467 + 8.5)/467 = 1.0182. The length of the upper edge will therefore be 21.5 ft, and of the lower edge 22.3 ft, which gives a keystone effect of 0.8 ft or 9.5 in. This would probably be noticeable were it not concealed behind the black rectangular border surrounding the screen. Figure 9.18 shows the difference in shape between the true image and the picture projected at 20° onto a vertical screen according to the assumptions in this example.

Projection Systems Involving Plane Mirrors

In many applications where space is restricted, it is common to use plane mirrors to bend the light beam from a projector around corners. This is a perfectly satisfactory procedure provided a few pertinent facts are given proper consideration:

(a) A projected image will suffer from *left-right reversal* after an odd number of reflections. It will also have left-right reversal if projected onto a translucent screen and viewed from the other side. Hence, if it is important to maintain a right-handed screen image, there must be an even number of mirrors for an opaque screen, or an odd number of mirrors for a translucent screen. If for some reason it is desired to use the wrong number of mirrors, the picture can be corrected, for still slides or a 16mm silent film, by turning the slides or film around left-to-right before inserting them into the projector. However, an 8mm cine film, or a 16mm sound film, cannot be reversed in the projector, and it then becomes imperative to use the correct number of mirrors.

(b) The *plane of the image* must accurately coincide with the plane of the screen. If this condition is not met, the image will suffer from keystone distortion and part of the picture will be in-focus while other parts are out-of-focus. When several mirrors are used, particularly at angles other than 45° to the light beam, it is often a difficult matter to determine whether or not the axis of the projector is truly perpendicular to the surface of the screen. However, careful tests for keystone distortion will often reveal a tilted axis, and one or more of the mirrors can be tilted slightly to rectify it.

(c) The *quality of the mirrors* is important, for a slight departure from perfect flatness will show up as a proportionate loss in definition when the mirror is close to the projector and the diameter of the elementary light beam is large. However, in that case the mirror will be small and it is not difficult to keep it flat. On the other hand, for a mirror that is distant from the projector, a departure from perfect flatness will appear as an irregular distortion of the image without any serious loss of definition.

(d) A faint *double image* is sometimes seen when a thick back-surface mirror is used in a projection system. This second image arises at the transparent front surface and usually has an intensity equal to only about 5% of the main image, but in spite of this low brightness it is sometimes clearly visible. At large angles of incidence, and particularly near grazing incidence, the first-surface glass reflection may become as bright as the back-surface silver reflection, and then the only cure is to use a front-surface mirror. Modern aluminum-on-glass mirrors, especially when anodized, or protected by an overcoating of silicon monoxide or magnesium fluoride, are entirely satisfactory provided that reasonable care is taken to prevent them from becoming dirty.

Projectors with Automatic Focusing

Recently, some slide projectors have been equipped with an automatic focusing device in which a very oblique beam of light is reflected at glancing incidence from the face of the slide toward a pair of photocells. If the slide is in proper focus, the reflected light falls idly between the cells, but if the slide is too far forward or backward, or if it is buckled, then one or other photocell will be illuminated and the projection lens will be moved in or out until the reflected light falls once again between the two cells. Of course, the initial focus setting of the projector must be determined by the operator; the automatic device merely restores each successive slide to the selected initial position.

The Overhead Projector, or Vuegraph

In recent years so-called "overhead projectors" have become common in classrooms and lecture halls, particularly where a large, bright projected image is desired. The arrangement is indicated in Fig. 9.19. A small, bright bulb, such as a quartz iodine lamp, is mounted below the writing stage, and a magnified image of the filament is projected into the projection lens aperture by a plastic Fresnel condenser having very fine steps, located beneath the stage. This arrangement ensures that the picture or writing, which appears upright to the lecturer when facing the audience, also appears upright to the members of the audience sitting in front of the lecturer and facing the screen, an obviously desirable situation. If the Fresnel condenser is well designed and well made, the filament image will be small, and it will serve to reduce the effective aperture of the projection

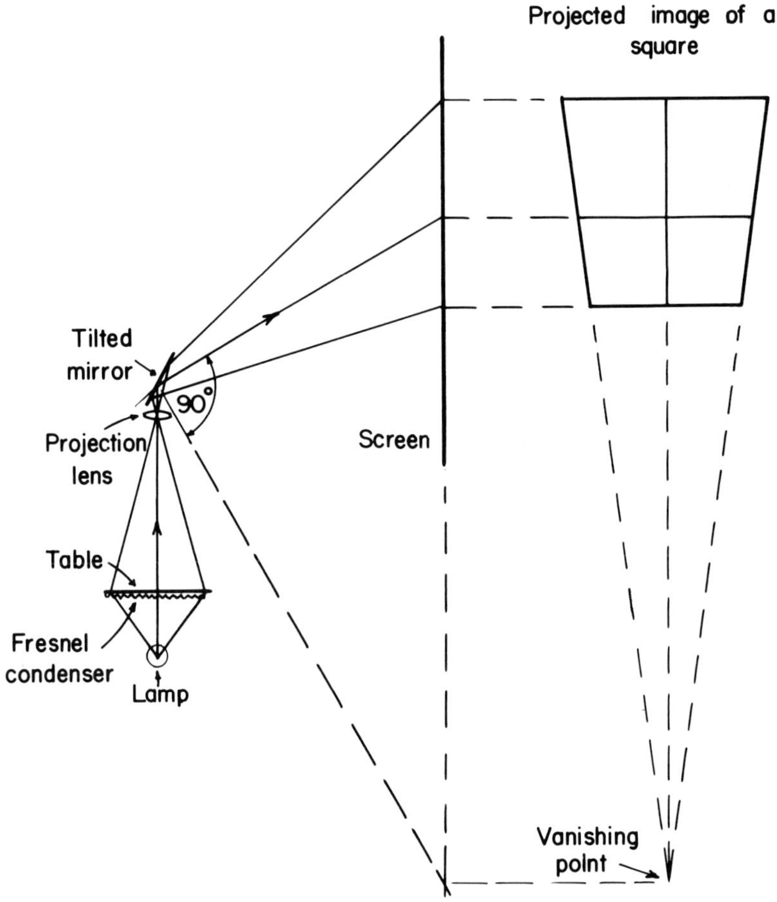

Figure 9.19. The optics of an overhead projector.

lens. If this aperture is *f*/15 or less, a very simple inexpensive type of projection lens can be used, such as a Periscopic objective consisting of two simple meniscus elements facing each other with their concave sides adjacent, the filament image falling midway between them. The steps in the Fresnel condenser must, of course, be so fine that their image on the screen is not visible to the audience, although they are sometimes quite noticeable to the lecturer.

Above the projection lens is mounted a plane mirror, nominally at 45°, to reflect the picture onto the screen, but this mirror can be deliberately tilted to raise the image on the screen and make it more readily visible to the audience. It is easy to see that this will introduce keystone distortion, the vanishing point being below the screen at the point where the screen plane intersects a line drawn through the mirror perpendicular to the axis of the imaging beam, as indicated in Fig. 9.19.

Care must be taken to ensure that the stage, which is often 8 × 8 inches in size, is fully illuminated, and that light from all parts of the stage enters the projection lens; otherwise, some parts of the picture will appear dark and poorly illuminated. In cleaning the surfaces of the plastic condenser, the greatest care must be taken to avoid scratches, as these will appear in sharp focus on the projected image.

Some small portable Vuegraphs have been made in which the light source is mounted close alongside the projection lens, a horizontal plane mirror below the writing stage serving to autocollimate the light and form the usual image of the light source in the projection lens aperture. The Fresnel condenser must be thin and rather weak, being specially designed for this application. It is essential that the stage be as close as possible to the understage mirror to reduce the possibility of double images, one of which is caused by light descending from the lamp and the other by light ascending toward the projection lens. Double images are always seen if the picture or the operator's pointer is raised above the stage.

Chapter 10

Stereoscopic Photography

Fundamental Principles

When we look at any solid object or assemblage of objects using both eyes, each eye observes a slightly different aspect of the scene. Thus, the pictures projected on the retinas of our two eyes are dissimilar, but the visual mechanism in the brain combines the two dissimilar pictures into a single three-dimensional, or "plastic" impression. This phenomenon is known as *binocular stereoscopic vision*.

It was early realized that we may divide the process into two separate and distinct stages. We may first photograph the two dissimilar views by means of two cameras situated in the positions occupied by each of our two eyes, and we may then view the two resulting photographs with our two eyes in such a way that the photograph taken by the right-hand camera is viewed only by the right eye, and that taken by the left-hand camera is viewed only by the left eye. When this is done, provided certain necessary conditions are fulfilled, we perceive the same three-dimensional image that we saw when we looked at the original solid objects directly from the camera position.

When the conditions are correctly chosen, the resulting plastic image will appear to be an exact full-scale model of the original object, apparently situated at its original distance from the camera. We speak of this correct type of stereoscopic reproduction as *orthostereoscopy*.

Depth Sensation in Monocular Vision

It is well known that people with only one useful eye manage to estimate distances with surprising accuracy. This is due to a number of causes that are quite different from the stereoscopy obtained as a result of binocular vision. In monocular vision we have a consciousness of distance by mentally comparing the apparent size of an object with its known true size, or by the falling of shadows, or by parallactic displacements resulting from a movement of the observer, or by the fact that one object is known to lie in front of another, or by a blueness and haziness of the atmosphere veiling distant objects, or even by the effort of accommodation of the eye at near distances, etc. These phenomena will not be considered further here, except insofar as they facilitate stereoscopic vision.

Pseudoscopic Vision

If by accident or design the left eye sees the right picture and the right eye the left picture, then the relative distances of all object points will appear to be reversed. Thus, objects that actually lie close to the eye will appear to be distant from it and vice versa. This apparent turning inside-out of an object is known as *pseudoscopic vision*. However, if a pair of stereoscopic prints is correctly mounted on a card, left-eye view to the left and right-eye view to the right, and the whole card is rotated through 180° in its own plane, the object will appear inverted but the stereoscopic reproduction will be otherwise unaffected. If each separate picture is turned through 180° about its own axis, then a pseudoscopic view will be obtained.

As a matter of fact, it is often very hard to realize pseudoscopic vision of real objects because we cannot imagine the objects looking that way. We have difficulty picturing an inside-out face, for example, but we can imagine a hollow hemisphere marked with the usual geographical features shown on a terrestrial globe. We can never achieve a pseudoscopic reproduction of a steel girder bridge or other structure in which near members come in front of more distant members. Moreover, the other factors outlined above, which are involved in the monocular estimation of distance, all act to inhibit the sensation of pseudoscopic vision.

The Stereo Camera

A stereo camera for the photography of moderately distant objects consists merely of two complete cameras, side by side, built into a single housing. Many satisfying stereoscopic photographs have been taken by means of two box cameras simply tied together with string, the two shutters being operated simultaneously; or by taking two pictures in succession, one directly and the other after moving the camera sideways by the normal eye separation of about 2.5 inches. However, for objects that may move, such as people, trees, or animals, a single instantaneous photograph is necessary

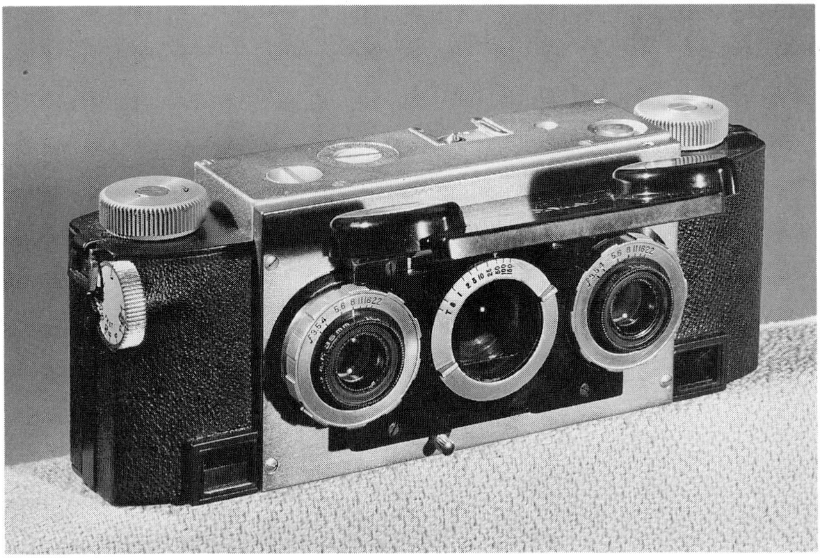

Figure 10.1. A typical modern stereo camera, the Stereo Realist.

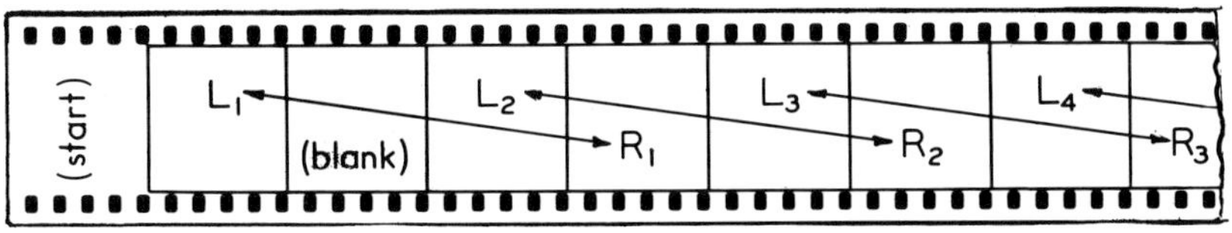

Figure 10.2. The picture arrangement in the Stereo Realist camera.

and some type of double camera with synchronized shutters must be used.*

The Stereo Realist (Fig. 10.1) is a typical American-made precision stereo camera using 35mm film, and each half of the camera makes a picture about 24 mm by 25 mm in size with a 35mm lens. The separation of the camera lenses is 71 mm, and the film winding mechanism is arranged to shift the film two 5-hole frames (47.5 mm) between exposures. The resulting picture arrangement is therefore that shown in Fig. 10.2, where L_1 and R_1 represent the left and right members of scene No. 1, and so on.

Necessity for Transposing the Stereoscopic Pictures

Suppose that Fig. 10.3 represents a stereoscopic camera, giving two in-

*A detailed description of many types of European stereo cameras may be found in the *Handbuch der wissenschaftlichen und angewandten Photographie*, Vol. 2, by K. Pritschow, Springer, Wien (1931); reprinted by Edwards Bros. Inc., Ann Arbor, Michigan. See also A. W. Judge, *Stereoscopic Photography*, London, Chapman and Hall (1950).

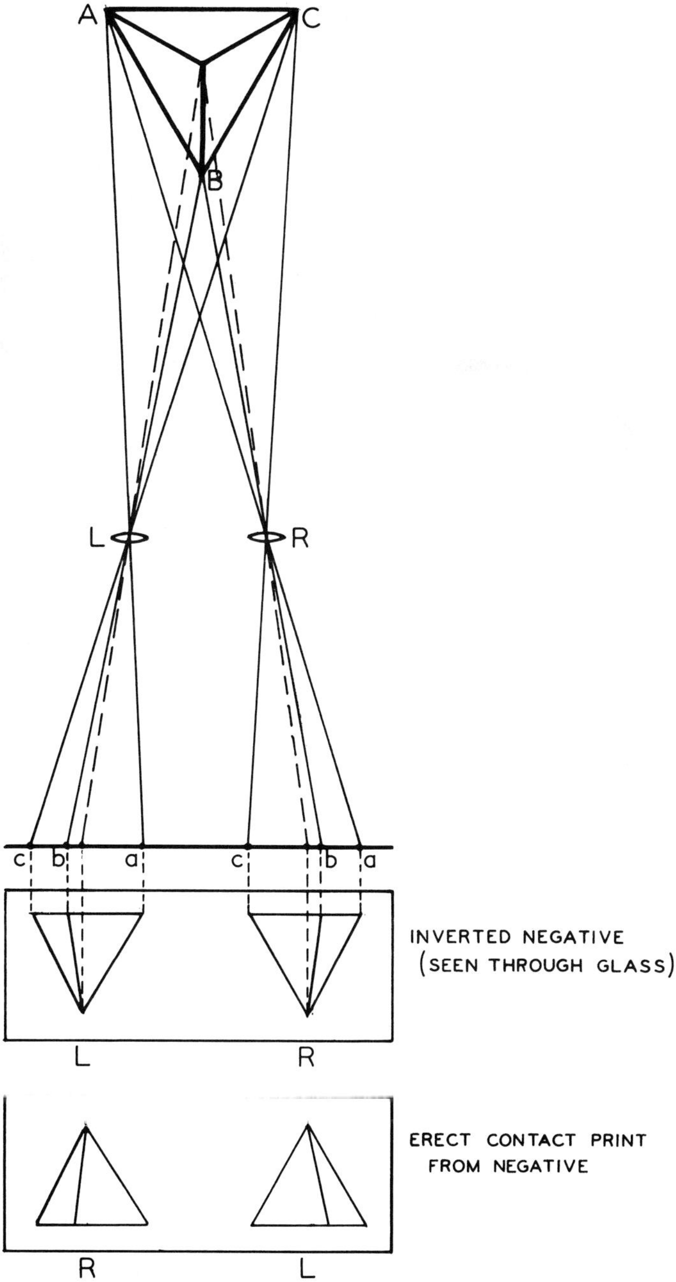

Figure 10.3. Illustrates why the prints must be transposed for stereoscopic viewing.

verted pictures side by side on a single negative. If the object is a pyramid with point up, for example, the negative images will, of course, be inverted with point down. If we now print this double negative in the ordinary way on a single sheet of paper, or on a glass or film transparency, the appearance of the positive print will be the same as the negative seen through the back, except for the photographic change from a negative to a positive. Thus, the positive print will be as shown at the bottom of Fig. 10.3 after it has been turned upside down to make the pyramid stand on its base correctly. It will now be clear that the view recorded by the left-hand camera lens has come to the right-hand side of the print, and vice versa. It is therefore necessary to cut the two prints apart and transpose them before mounting on a card for use in the stereoscope. What is really needed, of course, is for each picture to be rotated 180° *about its own axis* before printing. In mounting the two prints, their separation must be such that the two images of a distant object are separated by the same distance as they were before cutting.

Many mechanical devices have been constructed by which the two prints can be made, in succession, in their correct relative positions, such as by means of a special printing frame. It is also quite possible to take a

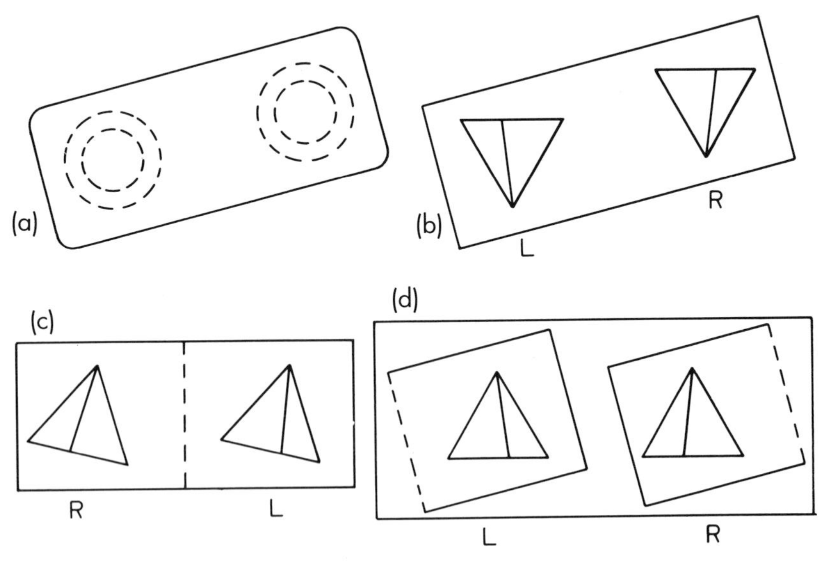

Figure 10.4. The effect of tilting a stereo camera sideways: (a) the camera seen from the back, (b) the negative in the camera, seen from the back, (c) the erect positive print, ready for cutting and transposing, and (d) the final prints mounted and ready for viewing in the stereoscope.

stereoscopic photograph of a stationary object with only one camera, by sliding the camera sideways by the correct amount after taking the first photograph and before taking the second. In some cameras the lens can be moved laterally, and the photographic plate moved simultaneously by twice as far in the same direction, so that the pictures will be already transposed on the negative.

Although it is obviously desirable to hold the camera level while taking the picture, a *small* sideways tilt will result only in a corresponding sideways tilt of the reproduced stereoscopic picture. If this is sufficiently great to be noticeable, it can be eliminated by careful masking of the mounted prints. The effect of a leaning camera is to rotate both pictures about their own centers through the same small angle (Fig. 10.4), and when mounting the prints, each must be rotated back again so that vertical lines in the pictures are accurately perpendicular to the base of the mounting card. The rules for correct perspective outlined in Chapter 1 become doubly important in stereoscopic photography.

Aerial Stereoscopic Photography

Stereoscopic effects may be obtained in aerial photography in two ways. The most common procedure, which is regularly used in aerial surveying, is to take two successive photographs of the scene separated by a brief time interval, the motion of the airplane thus producing a stereoscopic "base" between the two views. A slit camera (page 25) can be adapted to give stereoscopic views by using two lenses side by side in front of a single slit running across a broad band of film. Obviously, if the two lens axes lie in a plane parallel to the slit, both will give identical views of the ground, but if one lens is mounted slightly ahead of the other in the direction of travel, then the two views on the final film will be different. This is because the airplane will have moved between the time at which a given object is photographed by one lens and the time the same object will be photographed by the other lens, as shown in Fig. 10.5. In this diagram, the object A on the ground will be photographed by the leading lens when the airplane is at B, and by the trailing lens when the airplane is at C. The stereoscopic view will therefore be that seen by a pair of "eyes" located at B and C, with consequent great exaggeration of the depth effect.

The Stereoscope

The stereoscope for viewing a pair of stereoscopic prints consists ordinarily of two magnifying lenses, one for each eye, separated by the interpupillary distance of the average observer, which is about 63 mm (2.5 inches). Means

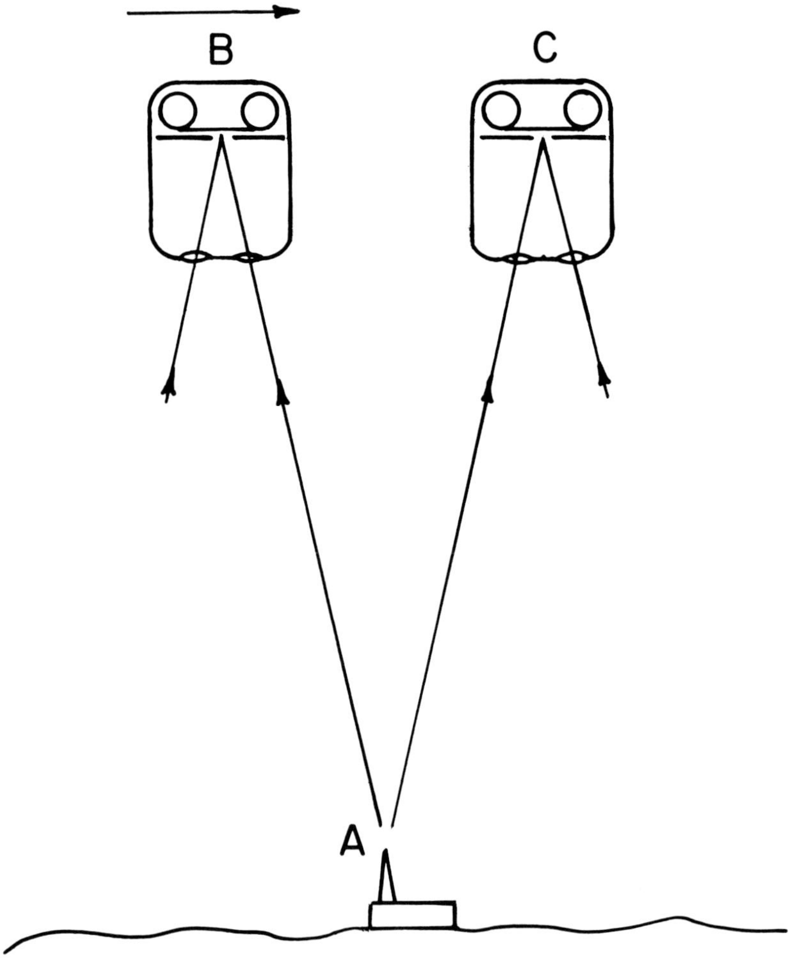

Figure 10.5. Continuous stereo strip photography from the air.

are then provided to ensure that the left eye sees only the left picture and the right eye sees only the right picture, the usual method being to provide an opaque septum between the two halves of the system. We shall see a correct orthostereoscopic reproduction of a group of *moderately distant* solid objects only if the focal length of the stereoscope lenses is equal to the focal length of the camera lenses, and if the separation of the camera lenses is equal to that of the viewing lenses. Failure to observe one or more of these conditions results in some form of distortion in the reconstructed plastic image:

(a) If the *focal length* of the viewing lenses is k times as great as the focal length of the camera lenses, all depth dimensions are increased by a factor k (including the depth of the reconstructed image), but the frontal dimensions remain unchanged. The effect is illustrated in Fig. 10.6(a) by a simple projection of fans of rays from the centers of the camera lenses and again from the centers of the viewing lenses. Even a 20% increase in the focal length of the viewing lenses can be readily detected. The effect is analogous to the familiar perspective distortion of a photograph when it is viewed from a point further away than the correct center of perspective.

(b) If the *separation* of the viewing lens is greater than the *separation* of the camera lenses, the effect is to reduce the frontal dimensions of the reconstructed image and all depth dimensions, as shown in Fig. 10.6(b). There is no simple numerical relationship in this case, but a glance at the diagram will indicate what is happening. If the prints are mounted too close together, the effect is the same as if the viewing lenses were too far apart.

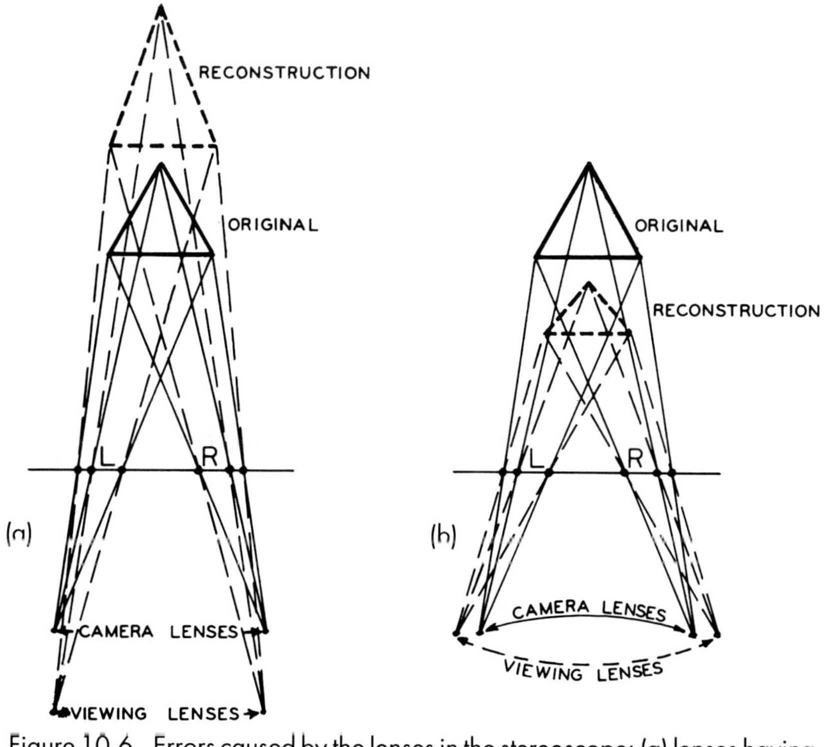

Figure 10.6. Errors caused by the lenses in the stereoscope: (a) lenses having too long a focal length, and (b) viewing lenses too far apart.

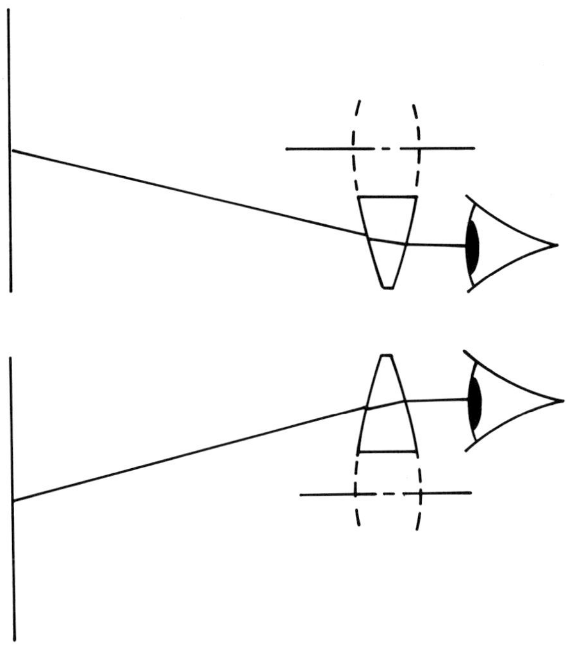

Figure 10.7. The Brewster stereoscope for viewing wide prints.

At one time it was common to mount the camera lenses further apart than the viewing lenses to obtain a deliberately enhanced depth effect. As the width of the separate prints was then equal to the distance between the camera lenses, they were too wide to be viewed in an ordinary stereoscope. This difficulty was overcome by Brewster, who suggested using somewhat decentered viewing lenses or a combination of a lens and a small-angled prism in the stereoscope (Fig. 10.7). The wedges are, of course, quite unnecessary if the proper conditions of orthostereoscopy have been observed.

The Transposing Viewer

If a photographic reversal or color process is used that yields small positive transparencies, it is often undesirable to cut and transpose the two halves of a stereoscopic pair before viewing. In such a case a *transposing viewer* may be used, one form of which comprises two parallel inverting microscopes having a very low magnifying power. The film is used in its inverted position with the left picture to the left and the right picture to the right, the two microscopes then serving to rotate the separate pictures through 180° about their own axes. In order to secure orthostereoscopic vision, it

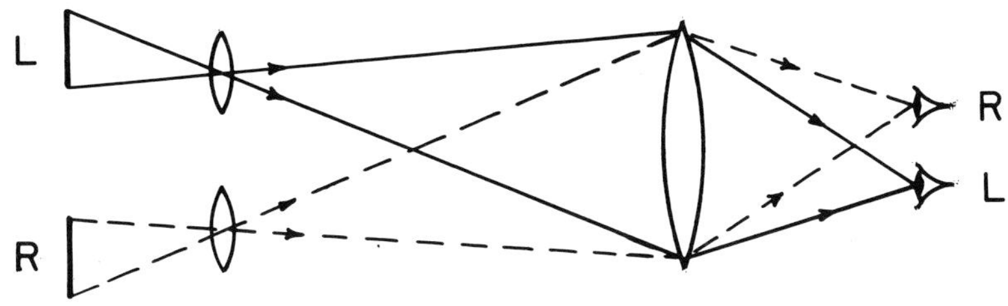

Figure 10.8. Stereo viewing in a large field lens.

is necessary that the equivalent focal length of each microscope be approximately equal to that of the camera lenses used to take the original photographs.

Stereo Viewing by Projection upon a Field Lens

A single observer can conveniently view a pair of stereoscopic transparencies by projecting them in register on a large field lens, the power of the lens being such as to form an image of each projection lens on the appropriate eye of the observer. A plan view of the arrangement is shown in Fig. 10.8.

The Mask

It is necessary to remember, when mounting a pair of stereoscopic prints, that the borders, or masks, bounding the separate prints are also viewed stereoscopically, and consequently form a window through which the observer imagines that he or she is looking at the reconstructed plastic image. For this effect to be satisfying the frame or window must appear closer to the eyes than the reconstructed image; the reverse situation is intolerable. This can be readily ensured by first framing one view with black paper for its pictorial effect, and then placing an identical mask over the other view, taking care that the lateral distance from the mask to the image of the *closest* object in the field is the same in both views.

Stereoscopic Photography of Very Close Objects

This case is much more complicated than the simple situation already discussed in which only moderately distant objects are involved, but by proper choice of the focal lengths, positions, and the separations of both the camera and viewing lenses, a true orthostereoscopic reproduction of real objects to any desired magnification can be obtained.*

*H. Kurtz, "Orthostereoscopy," J. Opt. Soc. Am. **27,** 323–339 (1937).

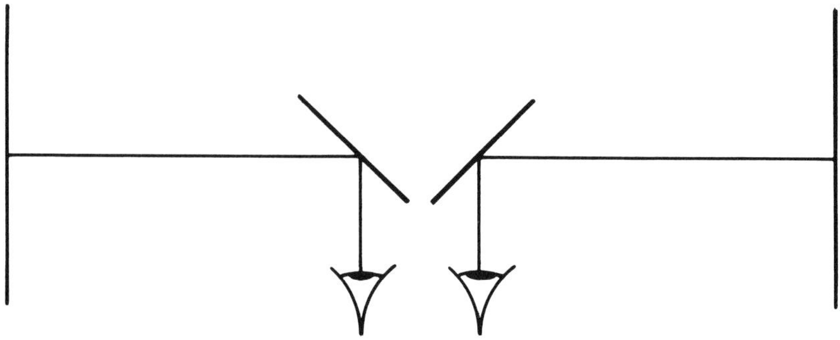

Figure 10.9. The Wheatstone mirror stereoscope.

When this is done, the final plastic image will be formed at a finite distance from the eye, and it is sometimes possible in such a case to choose the conditions so that the positive prints may be viewed without any lenses whatever. A convenient arrangement for this purpose is the Wheatstone stereoscope shown diagrammatically in Fig. 10.9. The device is used commonly for viewing stereo pairs of x rays to show depths in medical work.

Stereo Attachments

A single-lens camera may be used to form a stereoscopic pair of negatives by the use of a mirror or prism arrangement in front of the lens. The action of this system is illustrated in Fig. 10.10. It is clear that such an arrangement produces a pair of inverted images side by side on the same film, the left view being on the right and the right view on the left. When this film is printed and turned upside down, the two pictures come in the desired positions automatically without the necessity of cutting and transposing the prints. For many years the Leitz Company made an attachment of this kind for the Leica camera, known as the Stereoly, and a similar device is available for other cameras called the Stereo-Tach (Fig. 10.11). Suitable projection and viewing devices are also made to go with these attachments that enable each eye to see only its own picture at the correct magnification.

Colored Anaglyphs

An old and familiar device* for the convenient viewing of a stereoscopic pair of photographs without the use of lenses is to print the two images in

*This was apparently invented by Rollmann in 1856; it was later patented by du Hauron in 1891 (U.S. patent 544,666).

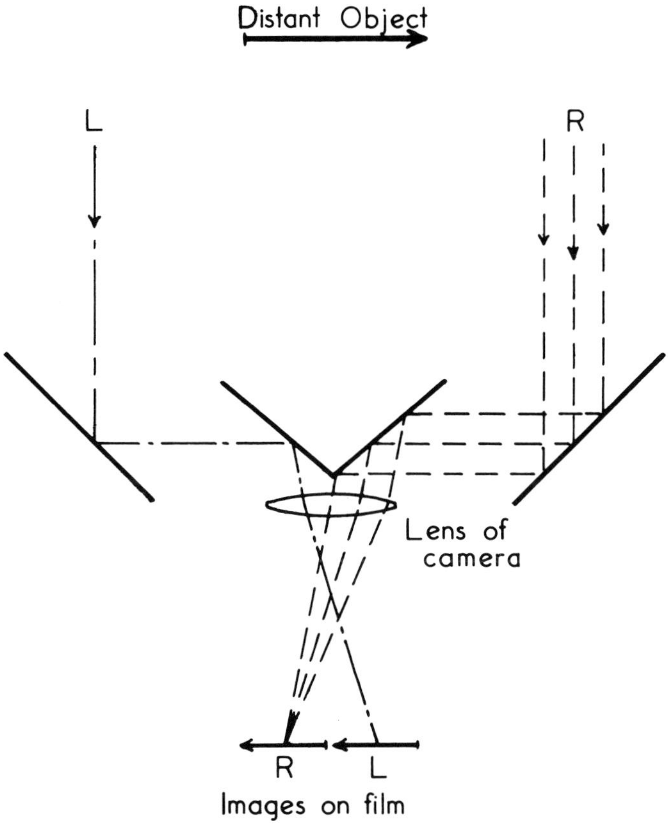

Figure 10.10. A stereo attachment.

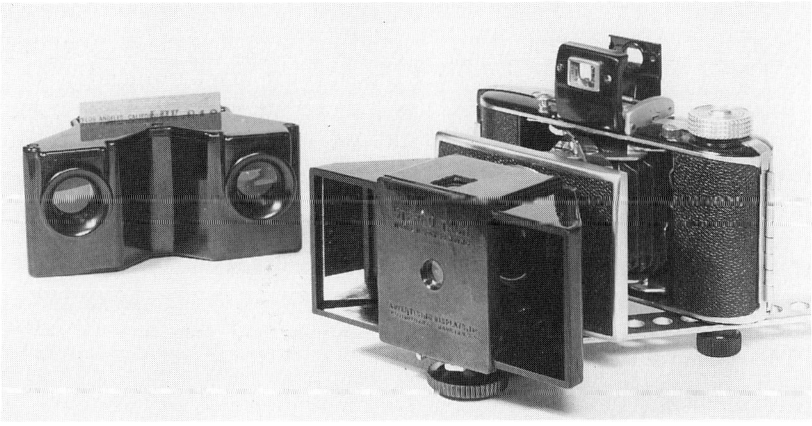

Figure 10.11. A typical stereo attachment and viewer.

approximate superposition in inks of complementary colors, say red and blue-green. The observer then looks at the picture through two gelatin filters of the same complementary colors, so that details printed in red, for example, will appear to be black when seen through the blue-green filter but will be indistinguishable from the paper itself when seen through the red filter and thus be invisible. Each eye, then, will be seeing only its own proper photograph, giving a stereoscopic effect in the usual way. The objections to this system are that only black-and-white pictures can be seen, the conflicting colors are a source of visual rivalry and cause fatigue, the inks and filters almost never form a perfectly complementary and mutually exclusive pair, and the necessary density of the filters gives a somewhat dim picture. Nevertheless, if well made, colored anaglyphs can be surprisingly satisfactory and many stereoscopic motion pictures have been based on this principle.

Stereoscopic Projection

A pair of stereoscopic transparencies can be readily projected in superposition on a screen, but is a matter of some difficulty to ensure that the observers' left eyes see only the left-hand view and their right eyes only the right-hand view. The principle of the colored anaglyph can be used; or the two projectors may be covered with polarizing filters, each member of the audience being equipped with complementary polarizers correctly oriented to separate the two pictures as desired. A metallized or other non-depolarizing screen must be used, and the common white diffusing or beaded screens are generally unsatisfactory.

Many mechanical devices were tried before sheet polarizers were available, but these all involved elaborate synchronizing means or accurately mounted mirrors, making them impractical for large audiences. The literature of stereoscopic motion pictures, and a survey of the patents, was explored by Professor E. J. Wall[a] with extraordinary thoroughness, and anyone interested in the history of this subject cannot do better than to refer to this article.

Many attempts have been made to devise means for separating the left- and right-hand views by means of a special ribbed screen and multiple projectors, so that any observer sitting in the correct position in the audience will see a stereoscopic reproduction without any spectacles or other viewing device. A pioneer in this field was H. E. Ives[b] but no really

[a] E. J. Wall, "Stereoscopic Cinematography," J.S.M.P.E. **10**, 326–344 (1927).
[b] H. E. Ives, "Motion Pictures in Relief," J. Opt. Soc. Am. **18**, 118–122 (1929); and **21**, 397–409 (1931).

practical means of stereoscopic projection of this kind has yet been devised.

In the mid 1950s there was a brief flurry of interest in 3-D motion pictures, which were shown in regular theaters and viewed through polarizing spectacles. However, the film producers and the audiences discovered quickly that good stereo motion pictures are very difficult to accomplish. The left and right films must be taken with two cameras mounted together with just the right separation and degree of convergence, and the result is satisfactory only when viewed from the correct position in the audience.[a]

One of the more obvious properties of projected stereo is that any point in the reconstructed three-dimensional image occupies a fixed fraction of the distance from observer to screen, and if the observer moves toward the screen, the depth of the image will become reduced although the frontal dimensions may remain the same.

It should be understood that there is a fundamental difference between viewing a paper print and viewing a projected image on a screen. In the former case, the empty paper is white and the bright areas of the stereoscopic images are ignored; it is only the *dark* parts of the pictures that are visible to the observer. In a colored anaglyph the dark areas in one picture are printed in red and appear black when viewed through a green filter, and vice versa. When looking at projected images, on the other hand, the empty screen is black and the eye sees only the *light* areas. The polarizing filters placed in front of the two projection lenses serve to polarize these light areas and make them selectively visible to only one of the observer's eyes; the black areas are not seen by either eye and are ignored.

Vectographs

This method of viewing a stereoscopic pair without the use of lenses was announced by E. H. Land[b] in 1940. The process depends on the production of an image in various *degrees of polarization* of light, instead of in various degrees of blackness as is the case in ordinary photographs. Such an image is invisible when viewed through a polarizer set parallel to the direction of vibration in the picture, but it achieves its full range of density and contrast if the polarizer is rotated through 90° so as to be in the "crossed" position.

---

[a] The geometry of stereo projection and the conditions that must exist in the camera, printer, and projector to give satisfactory pictures have been discussed in great detail in *Stereoscopic Transmission*, by R. and N. Spottiswoode, University of California Press, Berkeley and Los Angeles (1953).

[b] E. H. Land, J. Opt. Soc. Am. **30**, 230–238 (1940); also by B. Dudley in Photo Technique **3**, 30 (1941) and R. T. Kriebel in *The Complete Photographer* **9**, 3463–3466 (1943).

Land discovered that a partial polarizer can be made by combining a certain liquid reagent with a certain prepared plastic sheet, the degree of polarization being proportional to the quantity of reagent present. Suppose that an imbibition or wash-off-relief (dye-transfer) matrix is allowed to soak up the liquid reagent instead of a dye; then when it is squeegeed against a sheet of the prepared plastic the result will be a picture made up of degrees of polarization instead of degrees of density. It is then only necessary to look at the picture through a correctly oriented polarizer to see it in its full range of densities.

For stereoscopic purposes, one direct and one reversed matrix made from the original stereo negatives are used face to face in approximate register. Also, two prepared plastic sheets are bonded together back to back so that the direction of polarization in one is at 45° to the left of the vertical and in the other at 45° to the right of the vertical. The picture matrices are then soaked in the reagent, the prepared double plastic sheet is placed between them, and the whole sandwich is squeezed into contact in a small wringer. In less than a minute the reaction is complete, the matrices are peeled off, and after a brief wash the stereoscope transparency is ready for viewing.

The audience is equipped with polarizing spectacles of which the left spectacle polarizes light at right angles to that of the left-hand view in the Vectograph, thus making only the left-hand view visible to the left eye. The right spectacle is, of course, similarly arranged so as to see only the right image. Vectographs can be made in the form of a transparency, for direct viewing or for projection upon a metallized nondepolarizing screen, or they may be backed up with a metallic paint for direct viewing by reflected light. Vectographs are remarkably effective, and are accompanied by none of the defects that make colored anaglyphs unsatisfactory. It is even possible to make Vectographs in full natural colors.

It should be noted that if a Vectograph is rotated about its midpoint through 180°, a pseudoscopic view is obtained. This is because each view has now been separately rotated, the polarization conditions being unchanged by this rotation, so that each eye still sees the same picture as before.

Stereoscopic Pictures Requiring No Viewing Devices

Although it is not yet possible to view *projected* stereoscopic images without a viewing device before the eyes, this is entirely possible in a small single print or transparency. Two closely related methods have been used,

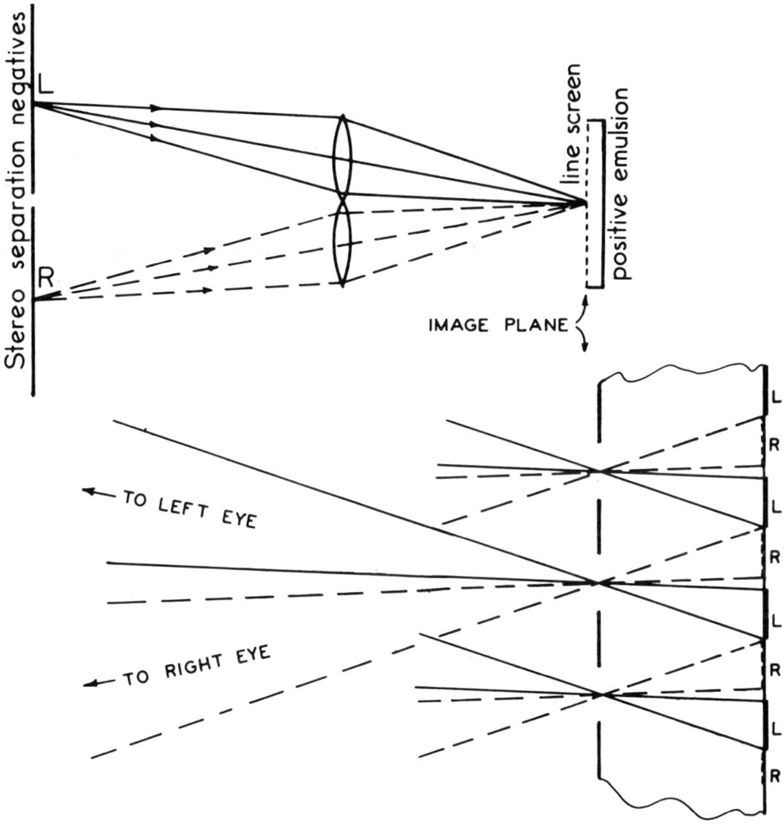

Figure 10.12. The line-screen method for direct stereoscopic viewing.

the so-called Parallax Panoramagram or Stereogram, which uses a parallel-line screen in front of the film, and the lenticular process ("Trivision" or "Vitavision"), in which the image is viewed through a ribbed structure molded on the back of the film base. These two methods have been actively developed for various purposes, both in color and black-and-white.*

The line-screen method is illustrated in Fig. 10.12. A pair of stereoscopic negatives is shown in position before two printing lenses, by which they are projected in superposition on a positive emulsion. In front of the emulsion and close to it is shown a fine-line ruled screen, such as a process

*L. P. Dudley, *Stereoptics*, Macdonald, London (1951); also T. Akoshi, *Three-dimensional Imaging Techniques*, Academic Press, New York (1976).

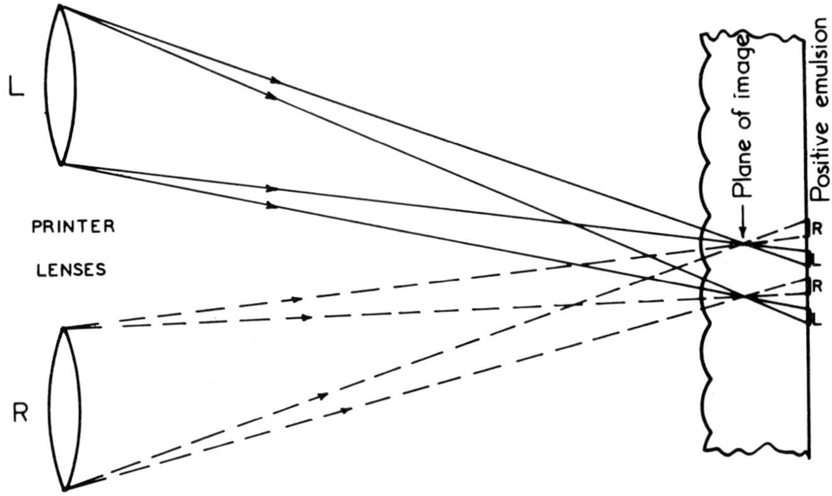

Figure 10.13. The lenticular method for direct stereoscopic viewing.

screen, with the rulings vertical. The two negatives are inverted and in precisely the same relative positions they occupied in the original stereo camera. The lines of the ruled screen are spaced apart at such a separation that the image elements projected by one lens on the film just fit in between the image elements projected by the other lens, as shown. If the positive transparency is now developed, fixed, and remounted behind the ruled screen, and if the observer's eyes are situated in the same positions as the two printer lenses, the left eye will see only the left picture elements and the right eye only the right picture elements. Provided the screen rulings are sufficiently fine, they will be scarcely visible, and each eye will see only its own apparently continuous picture.

In the other system, the back of the positive film is embossed with vertical lenticular ridges, which form a series of cylindrical lenses strong enough to project an image of the printing lens apertures upon the film. After the film is processed, the observer's eyes are placed in the positions occupied by the printing lenses, and the lenticulations then project back the separate image elements upon the eyes by merely reversing the direction of the light rays (Fig. 10.13).

This type of stereoscopic picture has also been made directly on a lenticular negative film by means of a large single lens. The advantage is that when the observer's eyes are moved from left to right, close objects appear to move across in front of more distant objects, and a much greater approach to three-dimensional realism is obtained. The process in this case is best understood by a consideration of Fig. 10.14, in which the large-lens

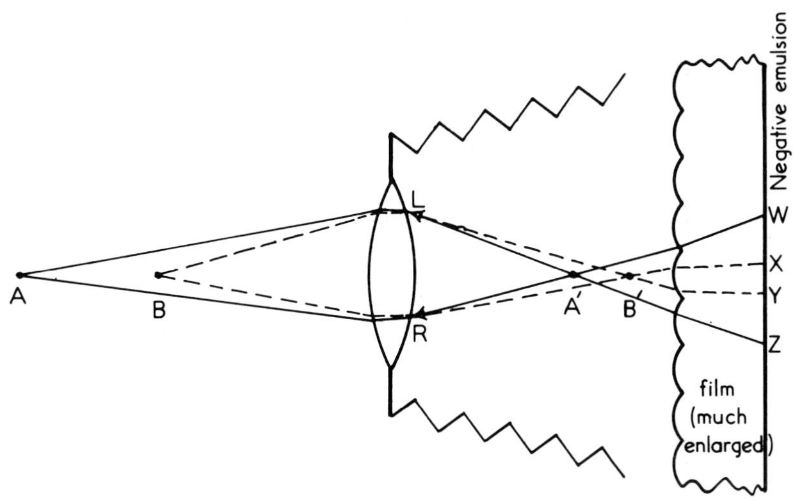

Figure 10.14. Making a lenticular stereo negative with a large lens.

camera is indicated, together with the assumed positions of the observer's eyes *L* and *R* in relation to the transparency. This last consideration is necessary because only those rays that enter the observer's eyes need to be considered. Suppose *A* and *B* are two object points at different distances from the lens. Their images will be formed at *A′* and *B′*. Then, so far as the useful rays through the right eye position *R* are concerned, the images of *A′* and *B′* will be projected by the lenticules onto the film at *W* and *X*, respectively. For the left eye, *A′* will be projected and recorded on the film at *Z*, and *B′* at *Y*. Thus, when both eyes, *R* and *L*, are used together, the images of *A* and *B* will appear to be reconstructed in space at their true positions *A'* and *B'*.

The lenticular film, therefore, serves as a storage device to hold three-dimensional images in their correct positions in space, even though these positions may be in front of or behind the film.

The original lenticular negative gives a pseudoscopic reproduction distorted in depth, for whereas *A* was originally more distant than *B*, it now appears closer to the observer, and since longitudinal magnification is equal to the square of the lateral magnification, the *depth* of the image will be quite out of proportion to its frontal dimensions. It is therefore necessary to print the negative onto a similar lenticular positive film in order to reverse the relative distances and re-form a correct stereoscopic positive transparency. This can be done either with a large lens similar to the original camera lens, or by a sliding lens arrangement in which the lens and unexposed positive film are made to slide during the printing exposure, in correct relative positions, in front of the illuminated negative. The distances and range of movements must of course be correctly worked out to produce an orthostereoscopic result.

It was pointed out by Bonnet in 1936 that the ruled or lenticular screen may be separate from but in contact with the negative film during exposure, and replaced into contact with the positive print after the processing and printing operations have been performed. Strict lateral registration of the ruled screen with the lines on the print is not essential, but the line *spacing* must be exactly right. This presents some difficulty with paper prints because of the expansion or contraction of the paper during processing, and as a result of subsequent changes in atmospheric humidity.

Multilens Stereo Cameras

If several small cameras are mounted closely together side by side in a compact row, each will form an image of external objects from a slightly differing viewpoint. The row of adjacent negatives can then be projected through a row of small lenses in such a way that all the images coincide upon a sheet of lenticular material backed up by a positive emulsion. It is necessary that each lenticule forms an image of all the projection lenses within the lenticular spacing. When such a lenticular picture is viewed directly, a stereoscopic image of the original objects is seen, and by slightly tilting the picture, the more distant objects appear to move laterally behind the nearer objects. The whole process has been discussed in detail by Okoshi.*

In 1981, a four-lens camera embodying this procedure was announced by Nimslo (Fig. 10.15). A brief description was given in the *British Journal of Photography*, Vol. 128, page 116 (1981). This camera contained four 30mm *f*/5.6 lenses, each of which formed an 18 × 21 mm image on two adjacent frames of ordinary 35mm perforated film. The row of negatives was sent to the manufacturer for printing on a special lenticular material having 150 lenticules per inch. Although some stereoscopic depth could be observed, there was a noticeable compression of depth in the scene, and the project was abandoned after a few years.

Binocular Magnifiers

A binocular magnifier (or "binocular loupe") is simply a magnifying lens that is large enough to allow both eyes to see through it simultaneously.

*T. Okoshi, *Three-dimensional Imaging Techniques*, p. 73, Academic Press, New York (1976).

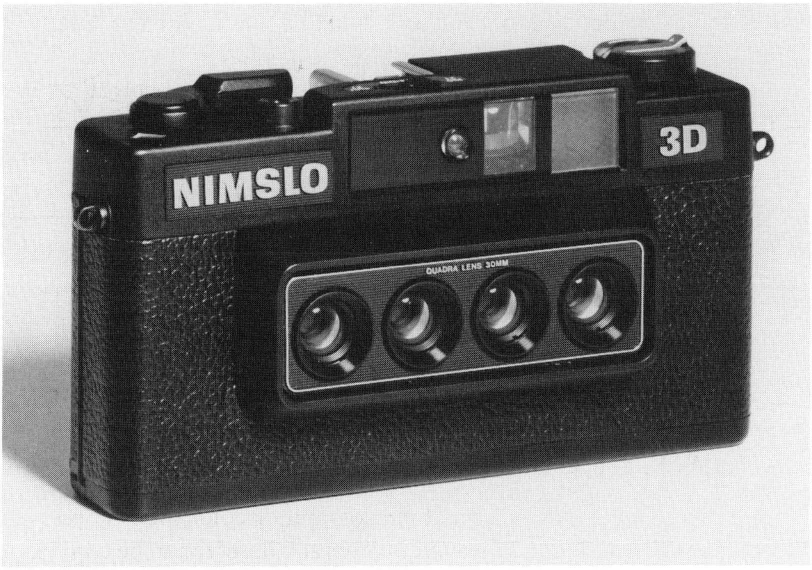

Figure 10.15. The Nimslo three-dimensional camera.

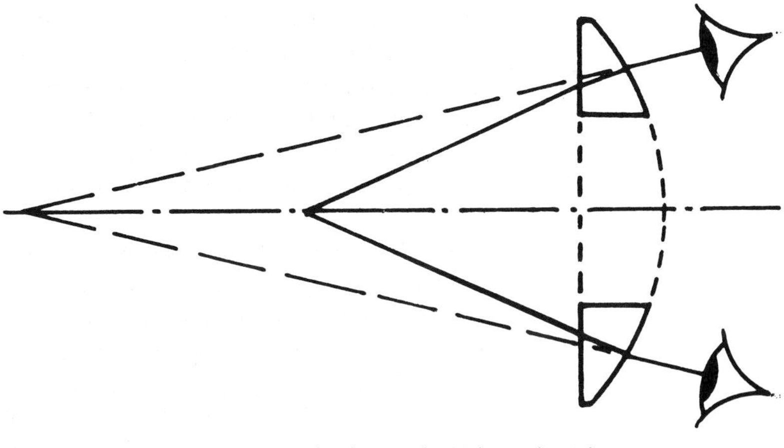

Figure 10.16. The binocular ("biocular') loupe.

Since much of the lens is not used, it is often cut down to a rectangular shape, and sometimes also the middle of the lens is removed, leaving two square or circular pieces for the two eyes (Fig. 10.16). It must be remembered that when using both eyes together, they must be converged to the image distance as well as be accommodated to that distance, so that no discomfort arises in observation of a single object through such a lens. A

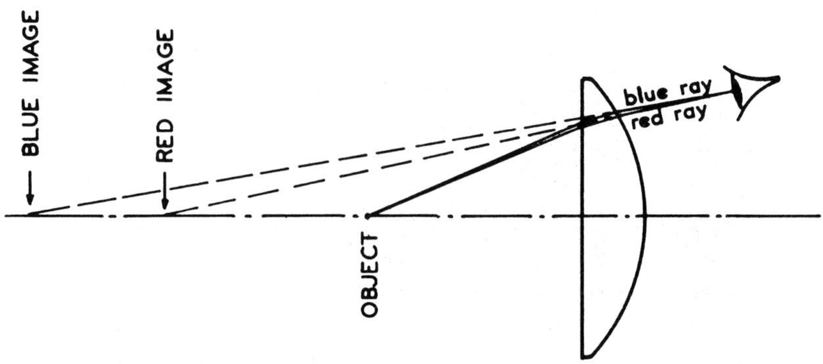

Figure 10.17. Red objects appear closer to the eye than blue objects.

lens large enough to be used with both eyes is now commonly said to be "biocular."

If a flat object such as a small photograph or color slide is observed through a simple lens in this way, it is generally observed to be convex in the up-and-down direction, but approximately flat, or slightly convex, in the left-to-right direction. The flat photograph thus appears to be a part of a cylinder with axis lying from left to right. If the lens is very strong, the cylinder becomes a torus, like the surface of a doughnut with its axis in the up-and-down direction. If the lens is tilted in any direction, the image appears to be tilted in the same direction but much more strongly; and if the photograph and lens are stationary and the observer moves about, the image will always appear to be square-on to his line of sight. These distortions are genuine stereoscopic phenomena, easily explained by careful ray tracing.

A marked stereoscopic color distortion is frequently observed if a strong single lens is used as a binocular magnifier in this way. Small red objects against a green or neutral area, for example, will appear to stand out closer to the observer, and similarly blue objects will appear to recede into the background. This is due to the dispersion of a ray passing through the outer portions of the lens (Fig. 10.17). This has the effect of forming the red image closer to the eyes than the green or blue images. The ordinary distortion of a simple lens used in this way is very marked, in the pincushion sense, but this is not usually a serious matter.

If binocular vision is desired without all of these stereoscopic distortions, a pair of simple magnifiers may be used with a system of mirrors and a beamsplitter, arranged so that each line of sight strikes the surface of the

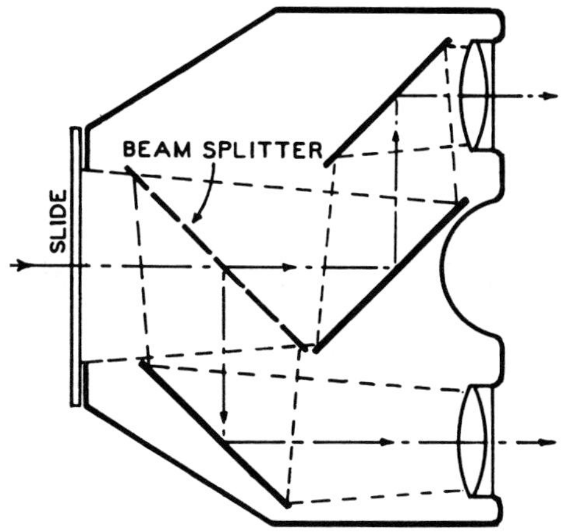

Figure 10.18. A binocular nonstereoscopic slide viewer.

photograph perpendicularly (Fig. 10.18). The difficulty of correctly aligning the four mirrors has prevented this type of slide viewer from becoming popular, although it can provide a very satisfactory binocular magnifier of about 4 inches focal length.

Chapter 11

Shutters and Flash

A shutter is an automatic mechanical device to open the light path through a camera for a predetermined and usually brief time, for the purpose of making a photographic exposure.

Most shutters also include a device to set the lens in the open position, either locking it open ("T" for time) or allowing it to remain open for as long as pressure is applied to the release lever ("B" for bulb, or nowadays, brief). In addition, all shutters provide for a very short "instantaneous" exposure, which may be of a fixed duration, say 1/25 or 1/40 second as in box cameras (Fig. 11.1), or varied at will from 2 seconds down to 1/1000 second or less in the more complex shutters (Fig. 11.2) or in focal-plane shutters.

An incredible amount of ingenuity has been applied to the mechanical design of shutters, from the simple flip-flop rotary shutters common in the earliest box cameras, through the more complex shutters equipped with flash synchronization that appeared after World War II, to the extremely accurate and reliable shutters made today. The timing of modern shutters is often electronically controlled. We shall not consider here the mechanical construction of shutters, but only those aspects that are of a purely optical character.

Types of Shutters

There are two unique planes in a camera system, the *aperture-stop* of the lens, and the *field-stop*, or film gate. Either of these planes is a suitable place for a shutter, and the two corresponding types of shutter, namely, the "between-the-lens" type and the "focal-plane" type, are well known. Each type has its

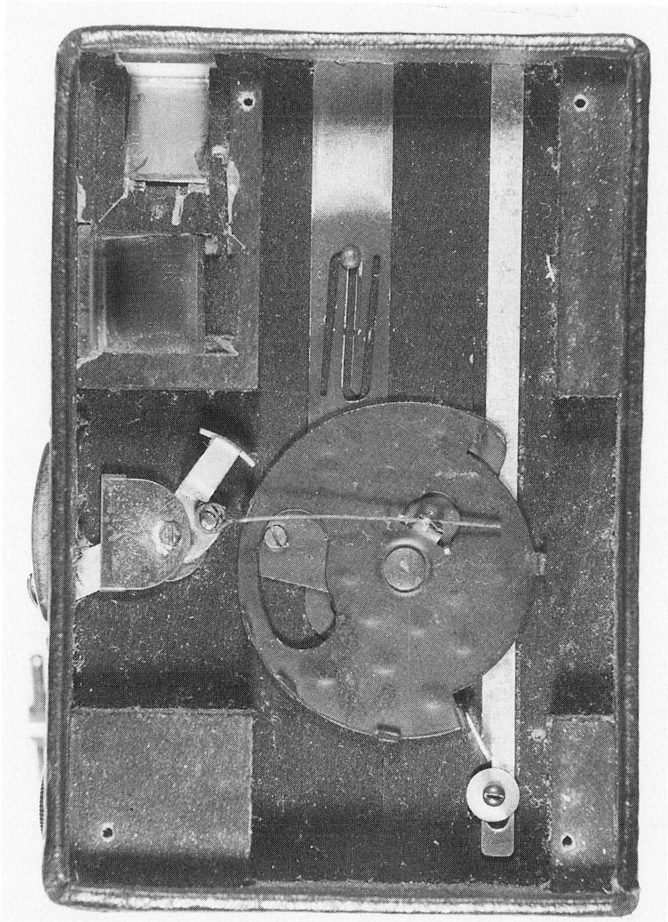

Figure 11.1. Simple rotary flip-flop shutter used on an early box camera.

advantages and disadvantages, which will be discussed briefly in this chapter.

Besides these normal types of shutter, there are some nonstandard or special types that are briefly mentioned here. Some cameras use a diaphragm shutter situated immediately behind the lens, to facilitate the use of interchangeable lenses. For aerial cameras, rotating disk shutters and louvre shutters are occasionally used. However, the placing of a shutter in any plane other than the two standard positions of at-the-lens or at-the-film is liable to lead either to low efficiency or to a nonuniform distribution of light over the camera field.

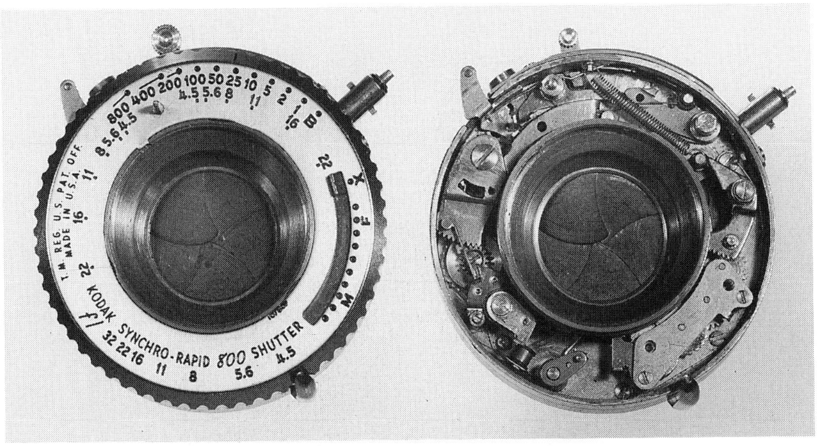

Figure 11.2. A modern shutter, the Kodak Synchro Rapid 800.

The Properties of a Between-Lens Shutter

A between-lens shutter is a device situated at or near the iris diaphragm of the lens,* which should open instantaneously to its maximum opening, remain open for the desired exposure time, and then close instantaneously again. Practical shutters, however, fail to perform in this idealized way because of the inertia of the moving parts and the delays inherent in mechanisms of this type. A graph of open area versus time of a typical between-lens shutter is indicated in Fig. 11.3. For the first few thousandths of a second (milliseconds) after the operation of the release lever, the shutter is opening up to its maximum aperture, then it stays open for a time, and finally closes again in about the same time that it took to open. The shaded area under the graph represents the product of illumination and time, which is a measure of the amount of light reaching the film and is the quantity that determines the density of the latent image on the negative.

Efficiency

The *effective* shutter time is the length of a rectangle, shown dotted on the graph in Fig. 11.3, that has the same height and the same area as the true shutter curve. It is thus the time that a perfect shutter would require to be open in order to pass the same quantity of light as is passed by the actual shutter. The *total* open time of the shutter is the time from the beginning

---

*There is no connection between the shutter and the iris diaphragm. They are generally mounted in the same housing for convenience.

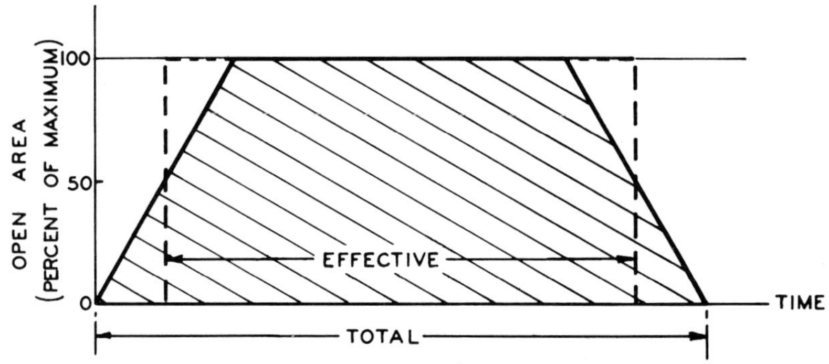

Figure 11.3. The performance curve of a between-lens shutter.

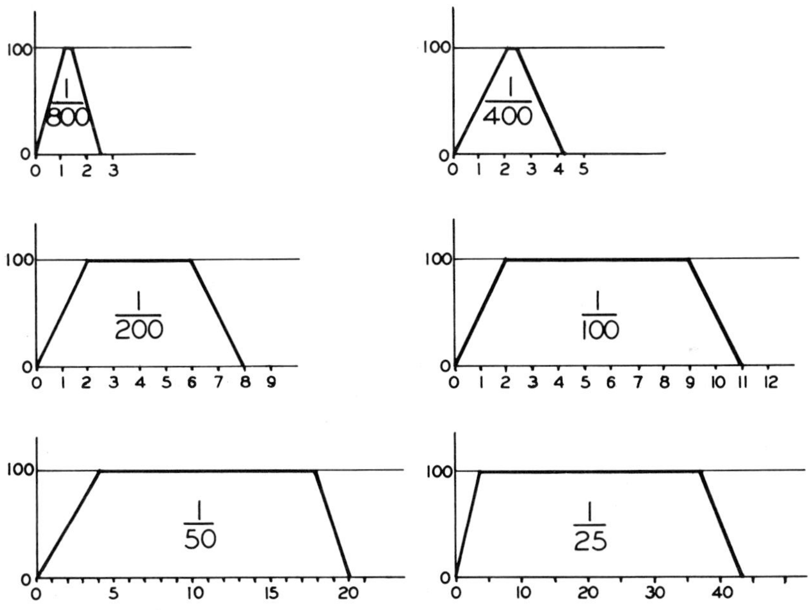

Figure 11.4. Performance curves of a modern shutter (in milliseconds).

of the opening portion of the curve to the end of the closing portion. The ratio of the effective time to the total open time is known as the *efficiency* of the shutter.

It is clear that for a very short exposure such as 1/800 second in which the shutter blades begin to close immediately after they have become fully open (Fig. 11.4), the efficiency will be about 0.55, or 55%, whereas for a longer exposure such as 1/5 second, the opening and closing times of, say, five milliseconds each are negligible in comparison to the full-open time of perhaps 190 milliseconds. In such a case the efficiency would be 195/200 = 97.5%.

It must be particularly emphasized that the *effective* shutter time, and hence the efficiency, varies with the diaphragm opening, whereas the *total* open time does not. The shutter manufacturer thus has a dilemma as to how the shutter speeds should be marked on the speed control dial or ring. There seem to be considerable arguments, however, in favor of marking the *effective shutter times at full aperture*. The reason for this is that the very short exposure times, where effective and total times differ most, are used only at or close to the maximum diaphragm opening. Then when the diaphragm is closed down, a longer exposure is given for which the question of effective versus total time is much less significant. By marking the shutter speeds in this way, therefore, the exposure will be correct for most applications. Indeed, most photographers never have any occasion to know the total open time of a shutter, but they are very much concerned with the effective exposure, especially with films of low latitude such as those in reversal color processes.

For aerial and other special cameras in which the shortest possible exposure time is necessary to "stop" motion, there is something to be said for marking shutters with the total time rather than the effective time,* but in ordinary photography the user has no means of knowing how short an exposure will be needed to make any particular moving subject appear stationary. In the paper cited, it is also pointed out that as the lens is stopped down one full stop, the efficiency of the shutter advances halfway to 100% from its original value.

Most shutters have two separate levers, one for cocking the shutter, and a release trigger that has to be moved only a short distance under light finger pressure to make the exposure. Sometimes one lever performs both functions, being raised for cocking and depressed to release the shutter. This presetting type of mechanism is likely to prove somewhat easier to hold steady than the "self-setting" or "self-cocking" kind common in lower-priced cameras. In this type a single lever must first be depressed to tighten a spring, the shutter being released automatically just before the lever reaches the limit of its travel. In many modern cameras the release lever is merely an electric contact requiring negligible finger pressure.

Some shutters are equipped with a "self-timer" or delayed action device, by which the release of the shutter can be delayed 10 or 15 seconds

*A. H. Katz, "Camera Shutters," J. Opt. Soc. Am. **39,** 1–21 (1949).

to allow the photographer to be included in the picture.

The Properties of a Focal-Plane Shutter

A focal-plane shutter consists of a curtain of cloth or other opaque flexible material that is made to pass rapidly across in front of the film. A slit of either fixed or variable width is provided in the curtain, allowing the exposure to be made (Fig. 11.5). In many early shutters the spring tension was adjustable to vary the linear speed of the curtain. In modern focal-plane shutters such as those almost universally installed in SLR and other 35mm cameras, one curtain starts to move across the opening and the second curtain is released a short time later so as to form a slit between the

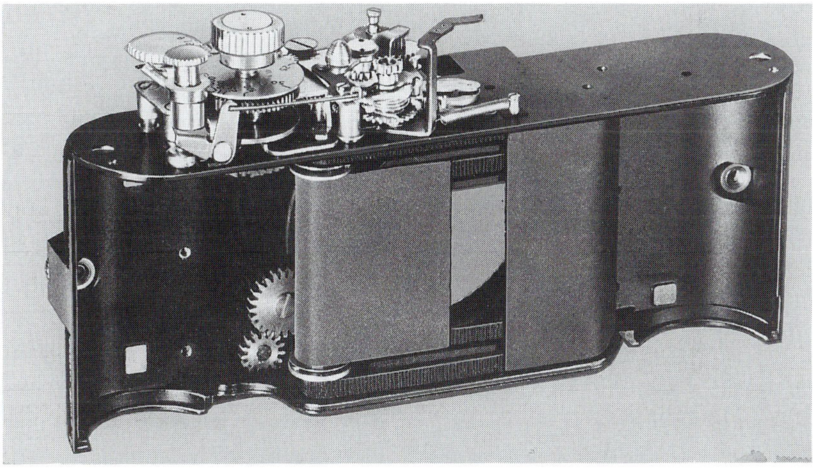

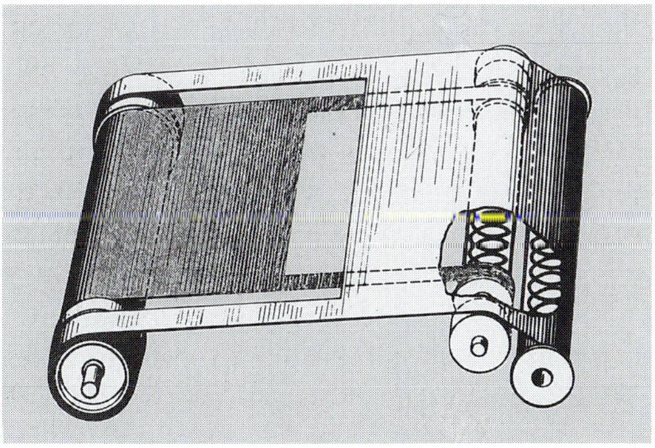

Figure 11.5. A typical small focal-plane shutter. The lower view shows the arrangement of the two curtains.

curtains passing across the film. For very brief exposures of the order of 1/1000 second, the slit is extremely narrow, say less than 1/8 inch, and its width must be made to increase slightly as the curtains speed up in order to yield a uniform exposure. At 1/25 or 1/30 second the first curtain just reaches the other side of the film before the second is released, and this is the shortest exposure that should be used with a flash unit (unless a long-burning FP bulb is used). For exposures longer than 1/25 second, a spinning gear and pallet device comes into play to hold back the second curtain for the required time before releasing it. Recently, the focal-plane shutters on many SLR cameras have become electronic, with a photocell and integrated circuit serving to measure the light passing through the lens and determining when the second curtain should be released.

When cocking the shutter, in order to avoid fogging the film, the first curtain comes up to and slightly overlaps the second, and then both curtains move across the opening to their starting point ready for the next exposure. In some older cameras such as the Graflex, no capping blind was provided, but the 45° mirror effectively shielded the film while the shutter was being cocked.

If the slit width is less than the film size, the actual exposure time will be less than the time taken to cover the whole film, in the same proportion. Consequently, if an object such as an automobile is rapidly moving laterally while the shutter slit is moving downward, the wheels of the car will be recorded earlier than the car body, with the result that the car will appear to be leaning forward and the wheels will appear to be slightly elliptical. This effect is well illustrated in the classical photograph by Lartigue, reproduced in Fig. 1.24. If an object is moving rapidly in the same direction as the curtain, its image will be somewhat compressed in the direction of motion, which is not ordinarily noticed.

Efficiency

If the curtain were very close to the film, each point on the film would be exposed to the entire lens aperture during the entire exposure time, which would reproduce the ideal conditions cited above for a between-lens shutter. However, for mechanical reasons it is necessary to place the curtain a short distance ahead of the film, where the cone of rays from the lens has a finite diameter. In that case the curve of illumination versus time would be as shown in Fig. 11.6. It is made up of three parts, one where the leading slit edge is moving across the circular ray bundle, one where the lens appears fully open, and one where the second slit edge is moving across the ray bundle to close it. The effective and total times of exposure and the

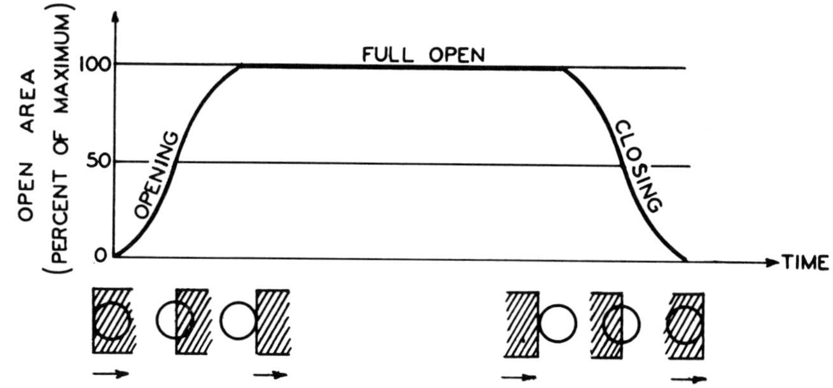

Figure 11.6. Performance curve of a focal-plane shutter.

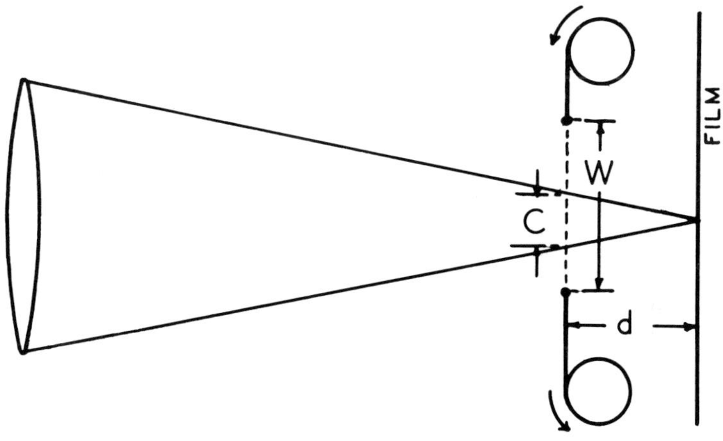

Figure 11.7. Efficiency of a focal-plane shutter.

shutter efficiency are defined in the same manner as for the between-lens shutter. However, the efficiency now depends only on the geometry of the situation and is given by

$$E = \frac{W}{W + C} , \qquad (11.1)$$

where W is the width of the curtain slit and C is the diameter of the cone of rays proceeding from the lens to a point in the image plane, measured in the plane of the curtain (Fig. 11.7). The quantity C is equal to d/N, where

d is the distance from film to curtain and N is the F-number of the lens. Thus, the greater d is made, the lower will be the efficiency of the shutter.

Since every possible ray from a point in the lens to a point in the film will be exposed for the same time, no matter where the curtain is situated, we are led to the conclusion that the *effective* exposure time, so far as negative density is concerned, is independent of the curtain position in the camera. However, as the curtain-to-film separation d is increased, the total time from the beginning of opening to the end of closing becomes greater, and the efficiency becomes correspondingly less. For "stopping" motion, therefore, the curtain should be very close to the film plane. For the photography of stationary objects, the curtain may be placed at *any* distance from the film without affecting the exposure.

Comparison of Between-Lens and Focal-Plane Shutters

In general we may draw the following conclusions as to the relative behavior of the two types of shutter:

(1) If the lens is small and the film large, then a between-lens shutter is the obvious choice, e.g., on a view camera. A large focal-plane shutter is difficult to make and is likely to be unreliable. Conversely, a between-lens shutter cannot be made larger than about 2 inches in aperture; hence, for a large lens a focal-plane shutter becomes a virtual necessity.

(2) A leaf shutter is likely to be better at the longer exposures, say over 1/200 second, whereas the advantages of a focal-plane shutter appear at the shorter exposures, say from 1/200 second down to 1/2000 second.

(3) For an SLR camera a focal-plane shutter is much more convenient, as it allows the shutter to remain closed during the setup and focusing operations.

(4) With a leaf shutter and interchangeable lenses, either each lens must be equipped with its own shutter, as on a view camera, or the shutter must be mounted close to the rear element of the lenses, which demands an unusual and often unsatisfactory lens design. If interchangeable lenses are to be used, then the focal-plane shutter is preferable.

(5) A focal-plane shutter restricts the choice of flash as the shutter must be fully open throughout the duration of the flash. Of course, an electronic flash makes everything much easier.

(6) There is little to choose between the two types of shutter in the matter of cost. They are both likely to be expensive if they are accurately designed and made.

Flash Photography

Open Flash

For a long time flash photography involved the use of an explosive mixture of magnesium powder with an oxidizer such as potassium chlorate. This was held in a shallow metal channel above the photographer's head, and ignited by a kind of cigarette-lighter arrangement. The operator opened the camera shutter immediately before firing the flash powder and closed it immediately after. This was known as an "open flash." (The cloud of smoke that followed was a most undesirable side effect).

Around 1930, lamp manufacturers began to offer a sealed glass bulb containing aluminum foil or wire in an atmosphere of oxygen, the metal being ignited by a small filament fired by a 3-volt battery. The open flash technique was still required, but the cloud of smoke was fortunately eliminated.

Flash Synchronization

Photographers strongly objected to the need for a separate opening and closing of the camera shutter, and as a result of united efforts on the part of the makers of flashbulbs and shutters it was found possible to couple a flashbulb to a small solenoid, which could then be used to release the shutter at the moment the lamp output reached its peak. In the early 1940s shutters began to appear in which no solenoid was needed, an internal timing mechanism being provided to give the correct delay between the firing of the flash and the opening of the shutter. Commonly this delay would be about 15 milliseconds, but in many of the better shutters an external lever was provided that could be set manually to any given desired delay between zero and about 30 milliseconds, to suit whatever flashbulb was being used.

For many years flashbulbs fell into four classes:

Class	Time to peak output (milliseconds)
X	zero
F (Fast)	5
M (Medium)	15 to 20
S (Slow)	30

A fifth group of flashbulbs were those labeled FP, intended for use on a camera having a focal-plane shutter. Because in this type of shutter a slit

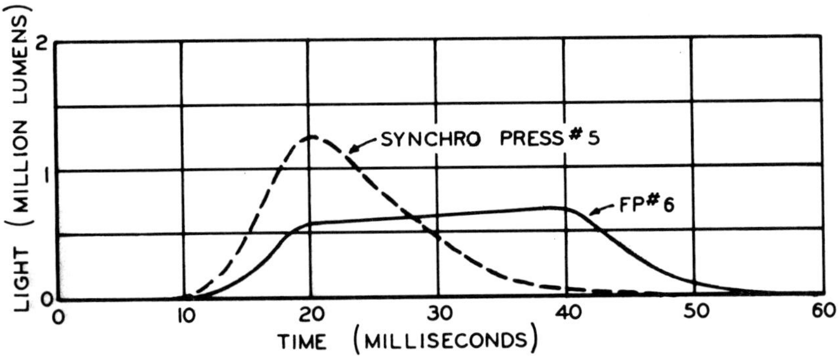

Figure 11.8. Comparison of an FP#6 bulb and a comparable normal flash bulb (Synchro Press No. 5).

of variable width is moved across the picture area, the flashbulb was required to burn at a fixed rate for at least 40 milliseconds (1/25 second). Comparable light output graphs for an ordinary flashbulb and an FP bulb are shown in Fig. 11.8.

During the period from 1940 to 1980 a wide range of flashbulbs was placed on the market by several manufacturers, notably General Electric, GTE-Sylvania, and Westinghouse. A listing of these can be found in each company's catalogs for that period. Most of the larger bulbs contained aluminum, but the smaller types often contained rare elements such as zirconium, hafnium, or rhenium. However, by 1990 electronic flash had become almost universal, the only "chemical" flashbulb still being manufactured being the AG-1B (the B refers to a blue coating to make the flash light resemble daylight).

Guide Numbers

As flashbulbs and film speeds vary over a wide range, photographers need guidance as to how to set the shutter for the best results. If the flashbulb is located at or near the camera, the illumination on the subject falls off as the square of the distance (the inverse square law). Hence, when the subject distance becomes greater, the lens aperture must be increased. However, since the exposure is inversely proportional to the square of the *F*-number, it turns out that the product of the subject distance and the *F*-number should remain constant. This constant is known as the *Guide Number*, and it varies with the type of flashbulb, the speed of the film, the size of the reflector (if any), and the shutter speed if the exposure time is less than the burning time of the bulb.

The Guide Number to be used under a variety of conditions is usually published in a table supplied with the flashbulbs. All the photographer has to do is to divide the Guide Number by the subject distance in feet to obtain the *F*-number that should be used on the camera.

Groups of Flashbulbs

A number of manufacturers have made plastic holders containing 4 or 10 small flashbulbs, so arranged that they can be fired automatically in succession. The well-known Flashcube contains four small bulbs resembling AG-1 lamps, each with its own built-in reflector, and so arranged that the cube is rotated 90° as the film is wound on to the next exposure. The Magicube is similar except that it is fired by an internal spring and requires no battery. The more recent Flashbar and Flipflash consist of a frame containing ten lamps that are fired in succession. The height of the Flipflash is useful as it tends to prevent the so-called "red-eye" effect caused by reflection from the subject's retina if the flashbulb is mounted too close to the camera lens. The Guide Number of the flashbulbs in these units is around 70.

Electronic Flash Tubes

An electronic flash unit consists of a sealed glass tube filled with xenon gas, plus traces of other gases to give a somewhat "warmer" light than the dead white of pure xenon. These tubes were first developed about 1935, their extensive application to photography being pioneered by Harold Edgerton. In 1941, his company, EG&G, made a large electronic flash unit that was marketed by Kodak under the name of Kodatron Speedlamp. This was large and heavy, and was suitable only for use in a studio.

By about 1960 several electric companies had begun to develop small battery-powered electronic flash units that were readily portable, and indeed were often built into the camera body. The operation of these units depends on the use of thyristors and other solid-state devices.* Direct current from a 6-volt dry battery is rendered intermittent and transformed up to about 350 volts, then rectified and used to charge the main capacitor. However, this voltage is too low to flash the tube. When all is ready, a small trigger capacitor emits a pulse that is transformed up to several thousand volts; this ionizes the xenon gas and permits the main capacitor to discharge through the tube. The duration of the flash is about a millisec-

*T. Karp, "The lowdown on portable electronic flash units," Mod. Phot., p. 82 (Jan. 1964).

ond, or even less depending on the exposure requirements.

The Guide Number of such a unit might be as follows:

Film speed, ASA	25	50	100	200	400
Guide Number	32	45	64	90	128

Note that the Guide Number is proportional to the square root of the film speed, all other factors being equal. After a few seconds delay the main capacitor will be recharged and ready for another flash.

A between-lens shutter must be set at X synchronization for use with the exceedingly brief electronic flash, and any shutter speed may be used. The problem of a focal-plane shutter is much more complicated. The shutter speed must be set at 1/30 second so that there will be an instant of time at which the picture area is fully exposed, and the flash must be timed to occur at that precise moment. Otherwise, part of the frame might be covered by one or the other of the shutter curtains.

The advantages of electronic flash over the simple chemical flashbulb are, of course, the possibility of making repeated flashes without having to replace the bulb, the very white color of the light, which closely resembles daylight, and the extremely short duration of the flash, making it an excellent motion stopper.

Shutter Speed Measurement

Many devices have been constructed for measuring shutter speed, some of which are simple and some very elaborate. We may, for instance, plot the actual curve of open area against time, either by the use of a photocell and cathode ray tube, or by recording an image of the shutter aperture on a moving photographic film. Alternatively, we may measure photometrically the integrated light passed by the shutter during the exposure, and thus obtain a direct measure of its effective speed. Or, we may measure the total open time by recording the light pulses from a moving flashing lamp on a fixed photographic film.

The measurement of shutter speed is likely to prove an attractive experiment for the serious amateur. Any moving object whose speed is known can be used for the purpose. For instance, one can move a small neon or argon lamp rapidly across the field of a camera while an assistant operates the shutter, and knowing that the lamp flashes 120 times a second, the total open time of the shutter can be determined by merely counting the flashes in the light trail on the film (Fig. 11.9). This method, however, is useful only for exposures longer than about 1/50 second or there will not

(a) 1 sec.	96 flashes	=	.800 sec.
(b) 1/2 sec.	53 flashes	=	.442 sec.
(c) 1/5	20 flashes	=	.167 sec.
(d) 1/10	10 flashes	=	.083 sec.
(e) 1/25	5 flashes	=	.042 sec.
(f) 1/50	3 flashes	=	.025 sec.
(g) 1/100	2 flashes	=	.017 sec.

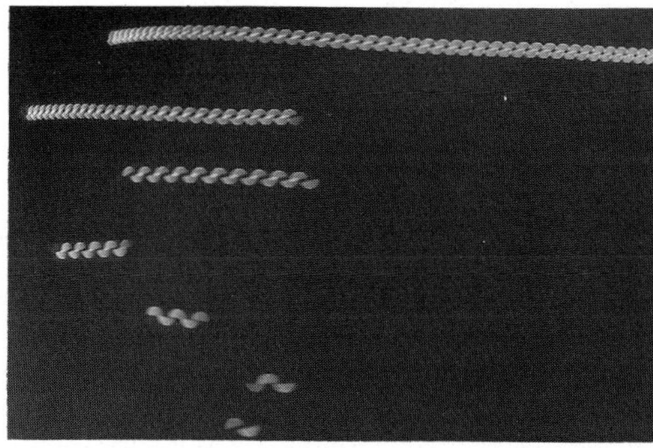

Figure 11.9. A typical film showing the 120-per-second flashes of a small argon lamp during seven shutter exposures.

be enough flashes to count. For the shorter exposures, a brightly illuminated white spot on the blade of an electric fan can be photographed, and the angular motion of the spot in its image on the film can be measured. This presupposes a knowledge of the number of revolutions per second of the fan. A similar device using a spinning bicycle wheel can easily be devised.

An interesting device for testing shutters in the laboratory has been described by Duffield and Lankes.* This method depends on placing a slit across the shutter aperture and projecting an image of it upon a disk of bromide paper on a rotating turntable, the slit image being radial to the disk. A sector-shaped image is formed on the bromide paper, the radial width at any instant being a measure of the diameter of the shutter opening at that instant. If the angular speed of the disk is known, it is possible to mark radial lines on the developed image at millisecond intervals, and thus the graph of shutter opening versus time can be readily plotted. A convenient speed for the disk is 600 revolutions per minute, or 10 per second, for then 1 millisecond corresponds to an angle of 3.6° on the disk. Furthermore, if a small neon lamp is flashed by the synchronizer contacts and also imaged on the disk, a measure of the flash synchronization delay time will be obtained. A focal plane shutter can be checked by this method, provided the shutter blind is moving in a direction radial to the disk.

*S. H. Duffield and L. R. Lankes, PSA Journal **15**, 85 (1949); See also Electronics, August 1948, p. 82.

Chapter 12

Camera Viewfinders and Rangefinders

The viewfinder on a camera is a simple but immensely useful accessory to picture taking, and an accurate viewfinder can help greatly in securing good composition. In particular, with color slides where the final frame is determined by the outline of the camera image and no cropping or masking is possible, careful use of the viewfinder becomes imperative. A precise viewfinder is also essential when taking motion pictures with a long-focus or telephoto lens; indeed, with a 6-inch lens on a 16mm camera, the best viewfinder would be a small telescope laid along and fastened to the camera body.

Most viewfinder masks are made slightly smaller than the full field recorded on the film to ensure that nothing of value in the picture will be cut off. For this reason it is generally desirable, and quite safe, to fill the viewfinder field with the desired subject matter in order that the picture should occupy as much of the film area as possible. In some so-called "sports finders" the actual picture area is marked by a rectangular line or other indication, the whole displayed scene being somewhat larger to give the user a better idea of how much of the scene will be photographed, especially with a moving subject as in sports events.

Types of Viewfinder

There are two main types of viewfinder, defined by how they are viewed. One type is called the *waist-level finder*, and the other the *eye-level finder*. Both types have been used extensively and are still in use. A further classification can be made depending on whether a ground glass is used to receive the image, or whether the system is specular with no diffusing surface. Of course, a tripod-mounted view camera having a full-size ground-glass screen does not require any type of viewfinder.

The Ground-Glass Viewfinder

The oldest type of viewfinder was itself a tiny camera, complete with lens and ground-glass screen, the image being erected and turned upward for convenient observation by means of an internal mirror (Fig. 12.1). The mask to limit the field was placed in contact with the ground-glass screen, and it often showed both vertical and horizontal picture areas if the finder could be hinged for use in either position of the camera. In box cameras, two separate viewfinders were often provided to make this hinged movement unnecessary. Great care had to be taken in manufacture to ensure that the image was accurately focused on the ground glass. The focal length of the viewfinder lens had to be kept short so that objects at all distances would appear in sharp focus, yet it had to be of large diameter to provide a bright image. The curvature of field of a simple lens, and the diffusion of

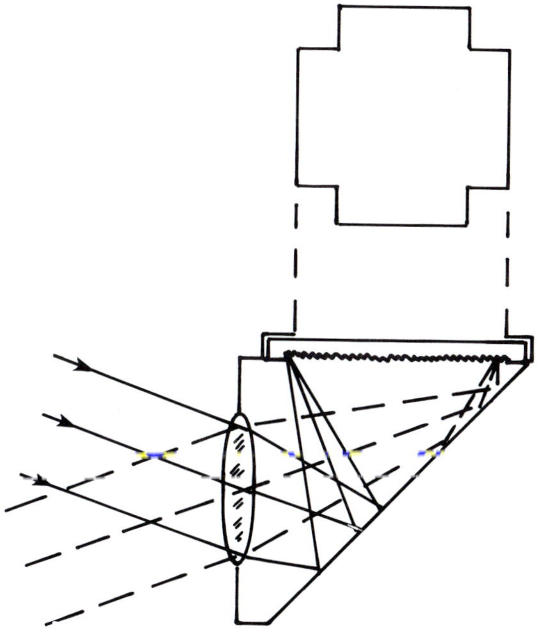

Figure 12.1. The ground-glass viewfinder.

ambient light by the ground glass, provided a definite limit to the usefulness of this simple type of finder.

The Twin-Lens Reflex Camera

The twin-lens reflex camera, which continues to be popular, may be regarded as two identical cameras built into one case, the upper having a 45° mirror and a horizontal ground-glass focusing screen, while the lower carries the shutter and film. Both cameras focus together so that the user can compose the picture and focus the image in the viewfinder and then take the picture when everything is ready [Fig. 12.2(b)]. Some of these cameras have a focusing front-board carrying both lenses, with a light-tight bellows or the equivalent to permit this mechanical movement. However, there is much to be said for the greater rigidity and accuracy of a solid

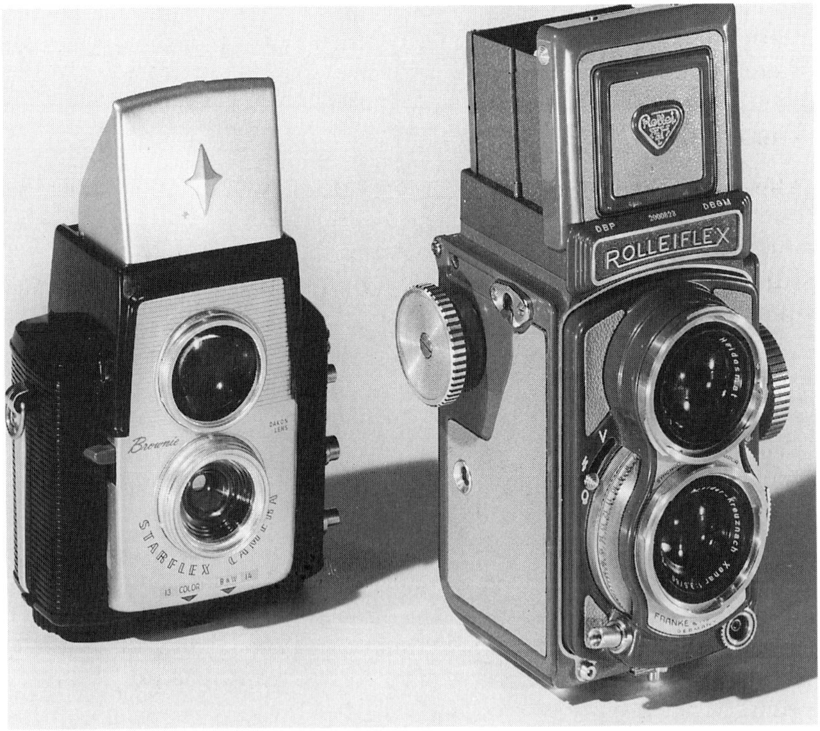

(a) (b)

Figure 12.2. Twin-lens reflex cameras: (a) with a large brilliant viewfinder mounted over a simple box camera, and (b) a camera with a ground-glass screen for focusing.

construction in which the lenses are focused by their front elements only, which are geared together for the purpose. In some reflex cameras, a field lens is mounted immediately above the ground-glass screen to help brighten the corners of the picture by directing the light toward the eye instead of relying solely on the scattering at the ground-glass surface.

A minor point, yet one that can be disturbing, is that in a twin-lens reflex camera there is no way by which the depth of field can be previewed. The upper viewfinder lens generally has a high aperture, $f/2.8$ or $f/3.5$, to assist in accurate focusing, while the lower camera lens can be stopped down to $f/16$ if desired. Thus, the depth of field in the photograph is likely to be much greater than the depth seen in the finder, but the photographer might not be aware of it and might not anticipate its effect on the picture.

A twin-lens reflex camera is normally used at waist level, but it can be converted to eye-level use by mounting a 45° mirror above the ground glass and adding a small magnifier lens in front of the eye to focus the image. The image will then appear upside down, but this should not bother the serious photographer.

The Brilliant Finder

A great improvement over the original ground-glass finder is the *brilliant finder*, in which the ground glass is replaced by a field lens of such power as to image the front objective lens aperture in the plane of the user's eyes (Fig. 12.3). The picture seen in this type of viewfinder is very bright, being almost as bright as the original object, and there is no scattering of ambient light as in the ground-glass type. Moreover, no particular care need be exercised by the manufacturer to ensure accurate focusing, as there is no definite image plane in this type of finder. The diameter of the front objective lens must be large enough so that its projected image easily includes both the photographer's eyes without any need to center the head accurately within the image.

Many low-cost cameras have been constructed in which a large viewfinder of this type has been mounted above an ordinary camera for ease in composing the picture [Fig. 12.2(a)]. The structure of the front lens of a brilliant finder is not important, and it is generally made biconvex or plano-convex. The field lens, however, must be carefully designed to eliminate distortion, the optimum form being nearly equiconvex. The large square field lens used in this type of camera is sometimes made of plastic material to save weight and cost.

The image seen in any viewfinder equipped with a 45° mirror appears erect but laterally reversed, so that an object moving across the scene from

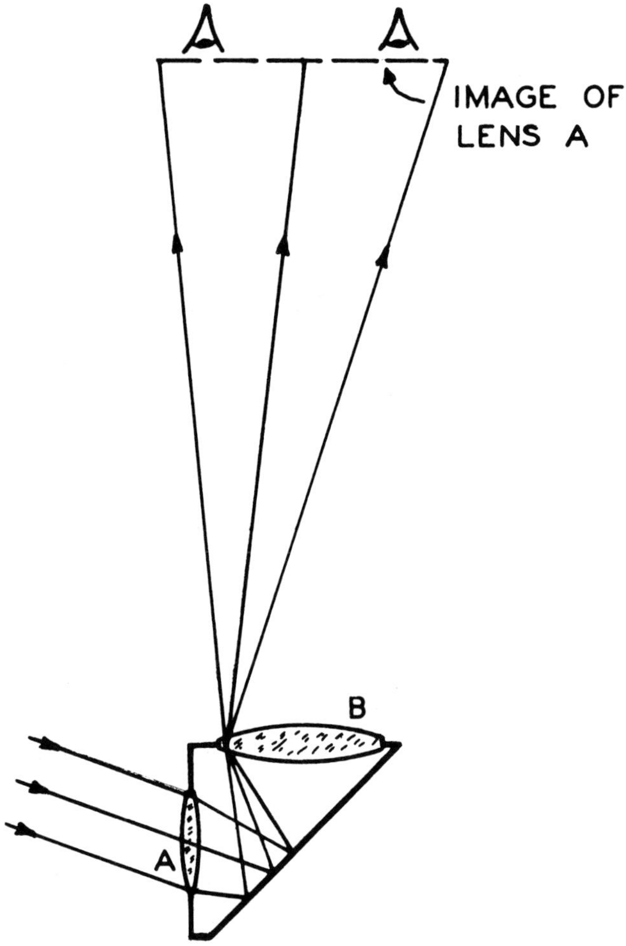

Figure 12.3. The brilliant viewfinder.

left to right appears to move right to left in the viewfinder. Reflex cameras of this kind generally have a square format to avoid the necessity of turning the camera on its side when photographing a tall object, as this would yield an upside-down image besides being awkward to manage.

Eye-level Finders

Since the 1920s there has been a growing demand for eye-level viewfinders to enable the photographer to clear crowds and other obstructions that may prevent waist-level viewing, and also to make the picture more closely resemble the scene as the photographer saw it.

The simplest finder of this type is merely an open frame to delimit the field and a small peephole at the back of the camera to look through (Fig 12.4). Because this arrangement is cumbersome and the delicate frame is liable to get damaged, and also because the eye is unable to explore as wide a field as the camera can record, the wire frame is usually replaced by a negative lens cut to represent the square or rectangular shape of the camera field. This lens forms a more easily observed reduced image of the scene; the small peephole is still required, however, to locate the eye on the axis of the lens. The mask surrounding the negative lens sometimes carries marks or notches indicating the upper limit of the field when the object is close to

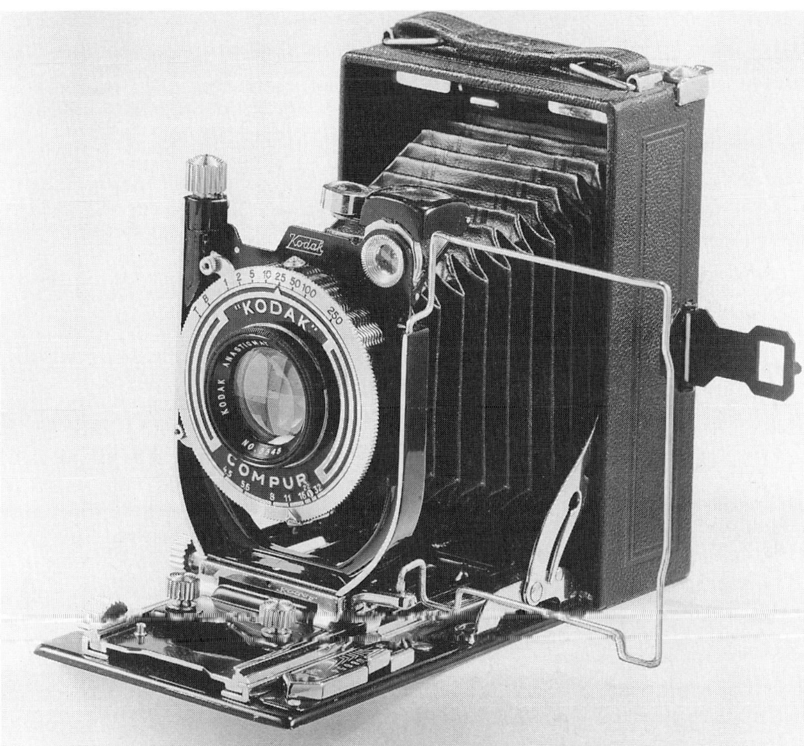

Figure 12.4. An open frame finder.

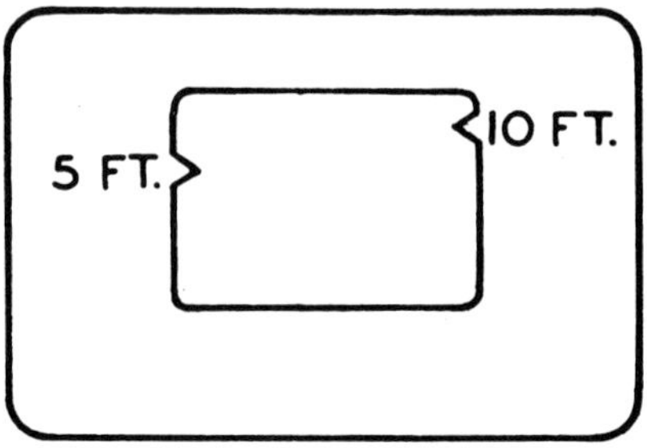

Figure 12.5. A Cine-Kodak viewfinder mask, showing the upper limit of the field for close object distances.

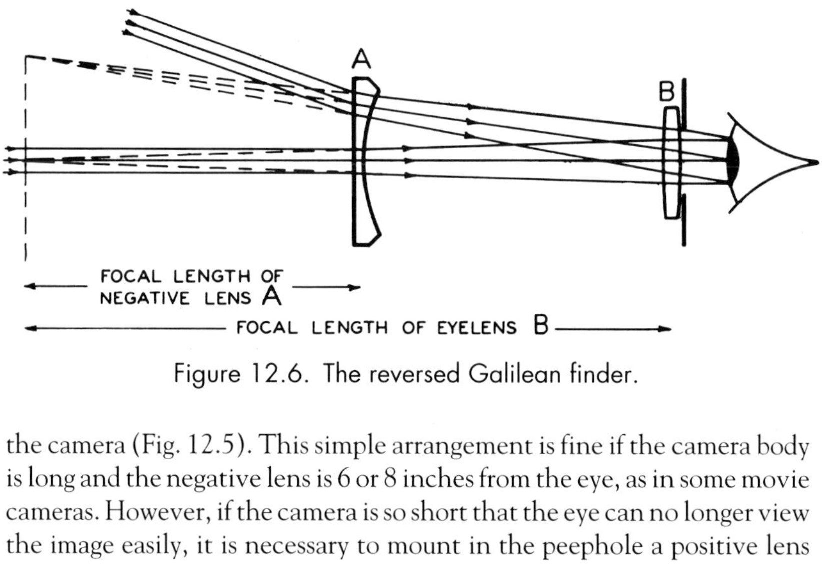

Figure 12.6. The reversed Galilean finder.

the camera (Fig. 12.5). This simple arrangement is fine if the camera body is long and the negative lens is 6 or 8 inches from the eye, as in some movie cameras. However, if the camera is so short that the eye can no longer view the image easily, it is necessary to mount in the peephole a positive lens focused on the image of the scene formed in the aperture of the negative lens. The whole arrangement then forms a reversed Galilean telescope having a magnifying power less than unity (Fig. 12.6).

Mask Parallax

Because the mask surrounding the negative lens is not in the same plane as the image of the scene, there will be a lateral movement of the scene within the mask if the observer's eye is moved sideways. This effect is known as *mask parallax*, and of course it spoils the accuracy of the viewfinder. It can be reduced by making the peephole very small, but a photographer wearing eyeglasses would then have difficulty seeing the entire scene. So the peephole must be large enough to accommodate the wearer of glasses, and then a serious amount of mask parallax will exist for those who do not.

Many inventors have tried to devise means for eliminating this mask parallax. One method is to provide a separate mask located at the anterior focus of the eyelens, through the use of a beamsplitter mounted between the observer and the large negative lens [Fig. 12.7(a)]. Another method is found in the Albada finder, in which the rear face of the negative lens is made partially reflecting to form a reflected image of the mask marked on the metal plate carrying the eyelens [Fig. 12.7(b)]. The dimensions of the system are so chosen that the mask image is formed at the same distance from the eyelens as the image of distant objects formed in the negative lens. Both systems suffer from the fact that in dim light the mask is hard to see, but of course the distant scene is then also dimly lit, so the problem is not really serious.

Mask parallax should not be confused with the type of parallax that arises with near objects due to the viewfinder being displaced laterally from the camera lens. This has already been referred to in connection with Fig. 12.5. Some cameras have been equipped with a tilting device to enable the photographer to bring the axes of camera and viewfinder into coincidence

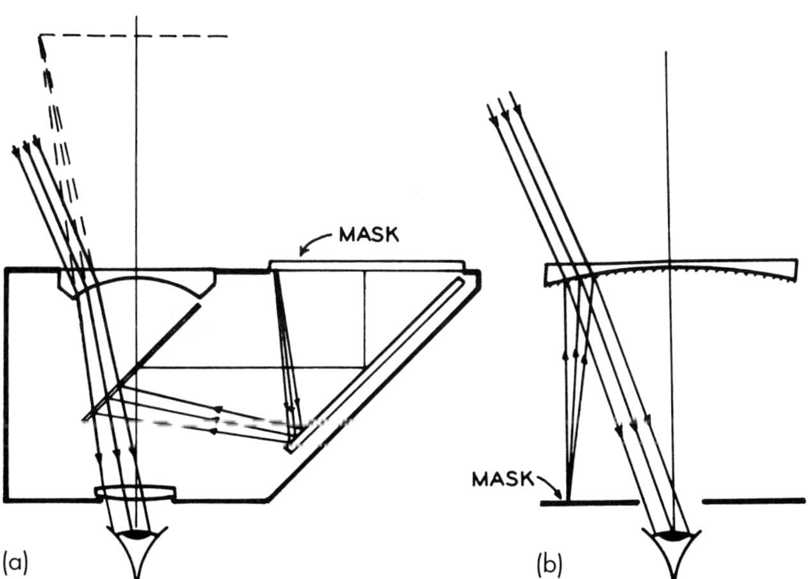

Figure 12.7. Two methods of eliminating mask parallax.

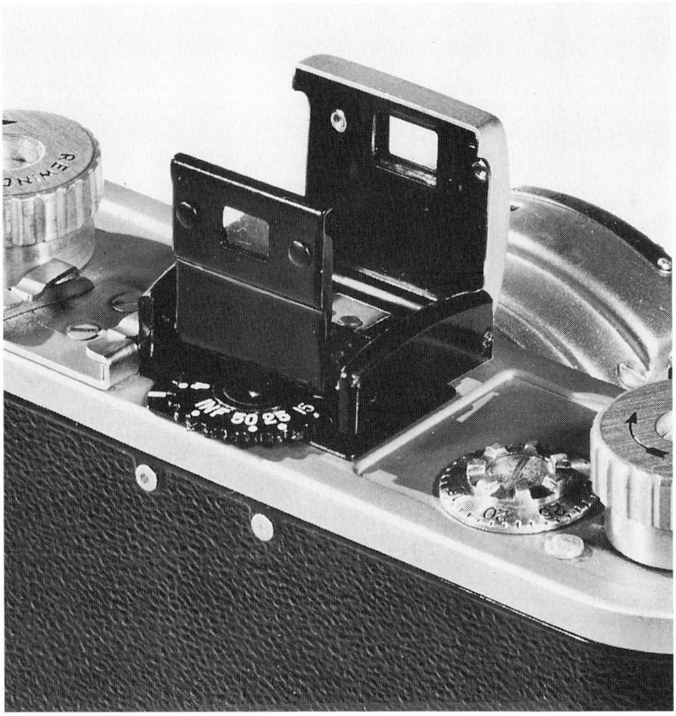

Figure 12.8. Viewfinder equipped with means to correct line-of-sight parallax.

at the actual distance of the object being photographed (Fig. 12.8). Of course, no parallax errors of any kind exist in an SLR camera.

The errors of a viewfinder may be classified as errors in *aiming* and *framing*. The parallax errors just considered are aiming errors; framing errors are much less liable to occur. These two types of error are basically independent. In more complex systems it is possible that the rim of the frame may be tilted relative to the edge of the film, which would constitute a rotational framing error. As has already been mentioned, it is customary to make the frame slightly smaller than the camera format to ensure that everything seen in the finder will appear on the photograph, but this allowance should not be greater than about 2 or 3% of the frame dimensions as it would lead to poor composition in color slides.

Magnification

The choice of magnifying power for a Galilean-type eye-level finder used on an ordinary still camera is a matter requiring some consideration. If the image appears very small to the eye, there may be difficulty in seeing the details of the subject, and everything looks so small as to be unreal. On the other hand, if the magnification is high, say 0.7× or more, the eye must scan the field to see everything, and as a result the user tends to concentrate on the principal subject and forget that everything in the viewfinder field will be recorded on the film. The net result is likely to be a picture in which the principal subject is small and surrounded by a wide expanse of unwanted material. Children often take pictures suffering from this defect because they tend to concentrate too much on the principal subject. An intermediate value of the magnification, say about 0.4×, is likely to prove best for most people. An open frame finder can be said to have a magnification of 1.0.

For motion-picture cameras, where the angular field is only about half that of a still camera, this trouble does not arise, and a magnification as high as 1.0 is quite acceptable.

Variable-Field Galilean Finders

In some movie cameras using interchangeable lenses, a sliding-lens arrangement has been provided to form a continuously variable-power telescope. The system used on the Magazine Cine-Kodak camera is shown in Fig. 12.9. Lens A and lens C are fixed, and the third lens *B* is free to slide along a track between lens A and lens C. With this simple arrangement, the system does not remain in afocal adjustment, but most users can accommodate their eyes to the small extent necessary to see the image. For use with the 15mm wide-angle camera lens, the front positive lens A can be swung down, while the sliding lens is set in the normal 25mm position.

A variable-field finder of this type has been used on some zoom cameras, the moving middle lens in the finder being directly coupled to the moving element in the zoom system. An alternative procedure that is generally much more satisfactory is to mount a partially reflecting mirror

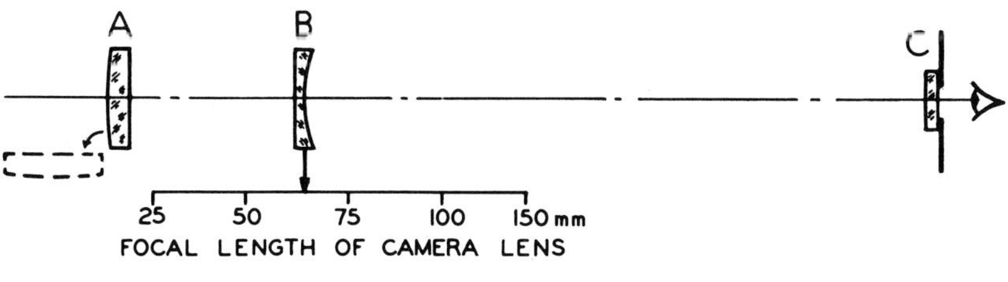

Figure 12.9. Variable-power finder used on the Magazine Cine-Kodak camera.

at 45° to the axis behind the zoom lens, to reflect a portion of the camera light to the operator's eye. The viewfinder system used in an early Canon zoom camera is shown in Fig. 7.21.

Telescopic Viewfinders

This name is generally applied to finders comprising a complete telescopic system with a positive objective lens, a lens or prism erector, and an eyepiece, the finder mask being placed in the plane of an internal image so that it is in perfect focus with no possibility of any mask parallax arising. This type of finder is ideal to use, but its cost has prevented its general adoption. A typical example is the supplementary variable-power "Universal Finder" made by Zeiss for the Contax camera (Fig. 12.10). In this finder five objective lenses of different power are provided on a turret, corresponding to the five focal lengths of the available camera lenses, the mask being permanently fixed in the body of the sight. Image erection is by a *K*-type roof prism situated between the two lenses of the eyepiece. A similar arrangement is used on some 16mm cameras, which have three lenses and three finder lenses on the same turret, so that when the turret is turned the camera and finder lenses are both changed simultaneously.

In some cameras a Fresnel lens is placed behind the ground glass to increase the brightness at the corners of the picture, and often a crossed-prism arrangement is mounted in the center as an aid to sharp focusing (Fig. 12.11). The effect of the crossed prisms is readily understood if we realize that an image lying in the plane of the exact crossing point of the prisms will not be affected by them, but if the image is formed above that

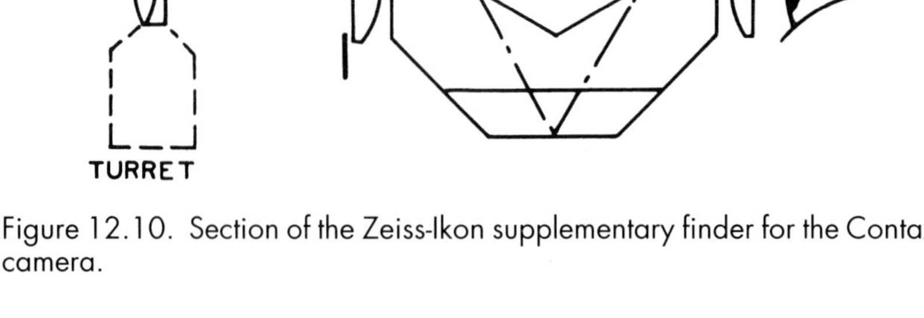

Figure 12.10. Section of the Zeiss-Ikon supplementary finder for the Contax camera.

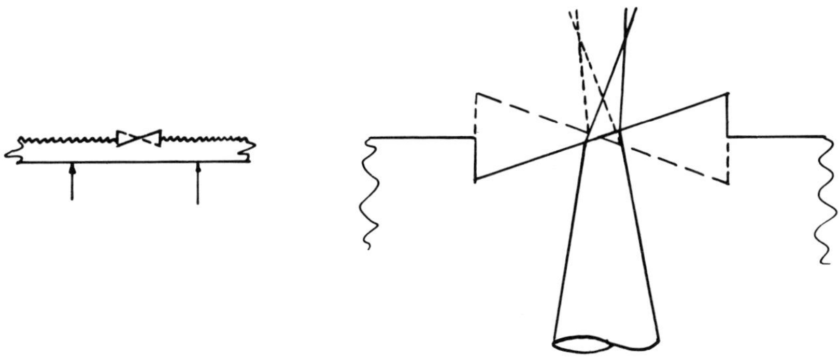

Figure 12.11. The crossed-wedge focusing device.

plane, it will be apparently divided, half being displaced to the right and half to the left. If the image is formed below the focused plane, the displacements are, of course, reversed. This simple device converts the photographer's blurred-or-sharp decision, typical of the ground glass, into a coincidence decision, which is much easier to make and far more accurate. This device is incorporated in the reflex finder system of the Canon zoom lens shown in Fig. 7.21.

Camera Rangefinders

The principle of the rangefinder is simple, although in many cases the actual apparatus required for its operation is elaborate and quite expensive to manufacture. However, in modern nonreflex cameras equipped with high-aperture lenses, it is almost impossible to guess the subject distance with sufficient accuracy, and some form of rangefinder becomes an absolute necessity. Indeed some photographers prefer a rangefinder camera over the much more convenient SLR for this very reason.

In Fig. 12.12, the object *A* whose range *R* is to be determined is viewed simultaneously from the two windows X and Y, the two views being brought together into the same eye by some suitable arrangement of mirrors. The eye thus sees two separate images of *A*, which may be brought into coincidence by one of several possible means, for example, rotating the mirror Y slightly. The amount of rotation necessary is proportional to the length of the base *B* between the windows. Rangefinders differ, then, in respect to (a) the type of coincidence to be observed, (b) the means adopted to deflect one beam to secure coincidence of the two images, and (c) the possible addition of lenses to produce a magnified image to aid in

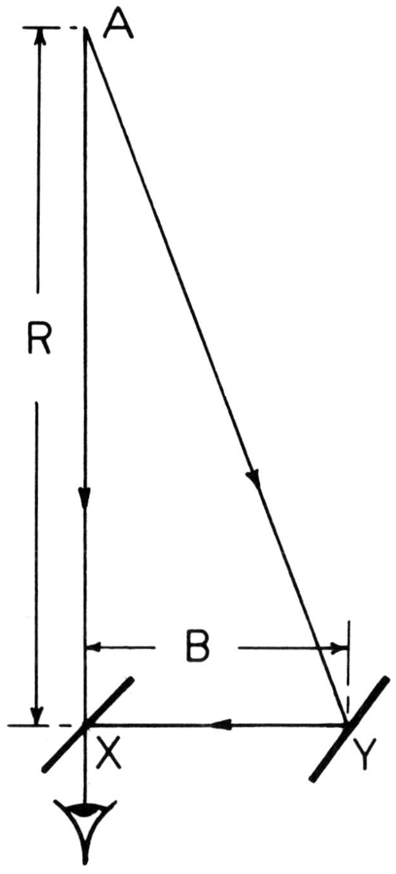

Figure 12.12. The action of a rangefinder.

making the coincidence setting accurately. The sensitivity of the rangefinder depends on the product of the base length and the magnifying power.

The Principal Types of Coincidence

In most camera rangefinders the exact point of coincidence is observed either by superposing the two images upon one another or by splitting the field along a horizontal dividing line with one of the two matching fields above it and the other below it. In addition to these two possible types, many large military rangefinders are stereoscopic, but to the writer's knowledge this principle has never been applied to a commercial camera rangefinder.

For the superposed double-image type of coincidence setting, the mirror X may be made partially reflecting and partially transmitting, so that about as much light reaches the eye from each of the two windows X and

Y. In this case both views of the object will be seen together, and it is a relatively simple matter to secure coincidence by rotating one mirror.

Alternatively, if the mirror X is made only half as high as the mirror Y, and if the eye is held back a little distance from the mirror X, the upper edge of this mirror will form a dividing line between the two fields of view, the lower half being that seen through the right-hand window Y and the upper half being that through the window X. Coincidence will now be observed by the upper and lower halves of the object appearing to fit together into one single continuous image. Range settings with a split-field rangefinder can obviously be made only on those portions of the subject that are at right angles to the plane of the triangle in Fig. 12.12; and if the object has only lines lying in one particular direction, it is necessary to tip the rangefinder into such a plane that this proper condition exists. Moreover, if the object is very small, it cannot be seen in both halves of the field simultaneously. For these two reasons, many users prefer the superposed double-image type of setting, although confusion is likely to arise in woodlands, or in a crowd of people, where many similar objects are distributed over the field and it is hard to be sure that the same object is being observed in both fields at the moment of coincidence. Sometimes lenses are added to increase the sensitivity of setting.

Coupled Rangefinders

Many cameras today are equipped with a rangefinder that is mechanically coupled to the focusing motion, so that the camera will be automatically in focus when the rangefinder fields are in coincidence (Fig. 12.13). This

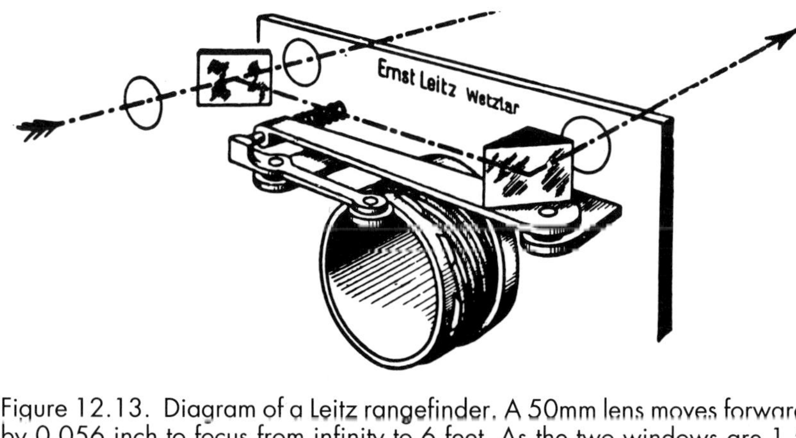

Figure 12.13. Diagram of a Leitz rangefinder. A 50mm lens moves forward by 0.056 inch to focus from infinity to 6 feet. As the two windows are 1.5 inches apart, the mirror must turn through 0.60° for a 6-foot distance. These data permit the design of the lever arms.

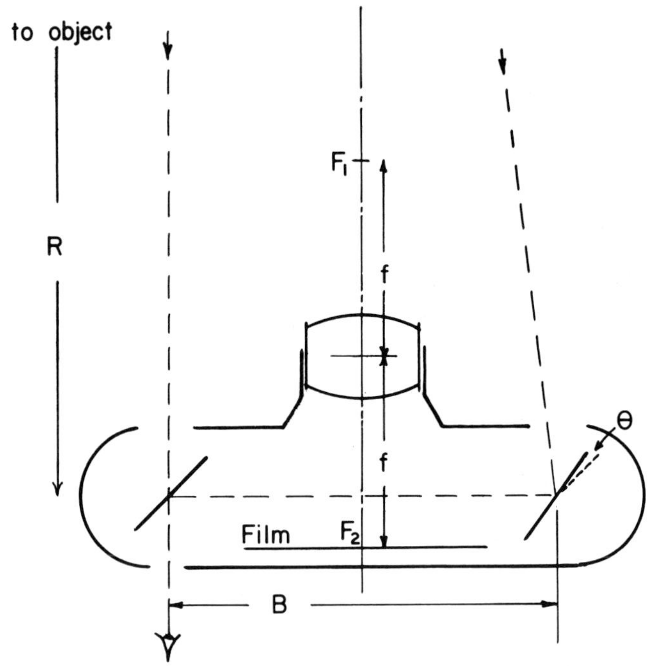

Figure 12.14. Diagram of a coupled rangefinder.

is obviously a great convenience to the user, but it presents a number of problems to the camera designer. The first problem is that the focusing motion of the lens is not linearly related to the tilting movement of the rangefinder mirror. The situation is indicated in Fig. 12.14. Suppose an object moves from ∞ up to a distance R from the plane of the rangefinder. The rangefinder mirror will need to be rotated through an angle $\theta = {}^1/_2\,(B/R)$, where B is the length of the base between the two rangefinder windows, whereas the lens must be moved forward through a distance x', which is equal to $f^2/(R-2f)$ (see formula on p. 32). As the object distance is diminished, the difference between $1/R$ and $1/(R-2f)$ will become progressively more significant. Many camera makers select a linkage to give the correct lens movement at one particular object distance and hope that the error will not be too obvious at other distances, but in the most expensive cameras this error is eliminated by the use of a cam. When a camera is equipped with interchangeable lenses that become automatically coupled to the rangefinder as soon as they are mounted on the camera, the cam must be part of the lens mount since its shape will vary with the focal length of the lens.

If the lens also carries a distance scale, as is generally the case, the manufacturer must correctly link three quantities, namely, the position of the image in relation to the film, the visible setting in the rangefinder field, and the reading on the focus scale. Moreover, these three quantities must be linked correctly at all object distances, which calls for considerable precision in manufacture and adjustment of the apparatus.

It might be thought that, because the accuracy of a rangefinder becomes progressively less as the object distance increases, a coupled rangefinder would become correspondingly less useful at great object distances. However, this is not the case, for the depth of field also increases with the object distance. A simple way of looking at this problem is to refer to the discussion of depth of field on page 87, where there is defined a "just acceptable angle of confusion" θ at the observer's eye that projects into the object plane as a small "circle of confusion" c. In a rangefinder, no observer is capable of making a perfect coincidence setting, and in general there will be a small uncertainty as to the degree of rotation of the mirror that is required to give apparent coincidence. This uncertainty causes a corresponding angular uncertainty as to the distance of the object, which corresponds exactly to the "angle of confusion" referred to above. Consequently, if the angular uncertainty of the coincidence setting in the rangefinder is less than the acceptable angle of confusion θ, the camera will be satisfactorily "in focus." This argument is independent of the object distance and consequently holds true for all object distances. This problem becomes negligible if the length of the rangefinder base is greater than the diameter of the lens.

Combined Rangefinders and Viewfinders

There is a great deal to be said in favor of combining the viewfinder and the rangefinder fields of a camera in one eyepiece, since the user is then freed from the necessity of quickly transferring attention from the rangefinder eyepiece to the viewfinder eyepiece immediately before taking the photograph. However, for a camera covering the normal 26° semi-field, some angular reduction is necessary in the viewfinder to enable the eye to take in the whole field at a glance, whereas in the rangefinder a magnifying telescopic system is generally needed to give sufficient accuracy. As these two requirements are incompatible, combined range and view finders are generally given a long base to compensate for the lack of magnifying power. A typical combined system incorporating an overall image reduction and a superposed double image in the central area of the viewfinder is shown

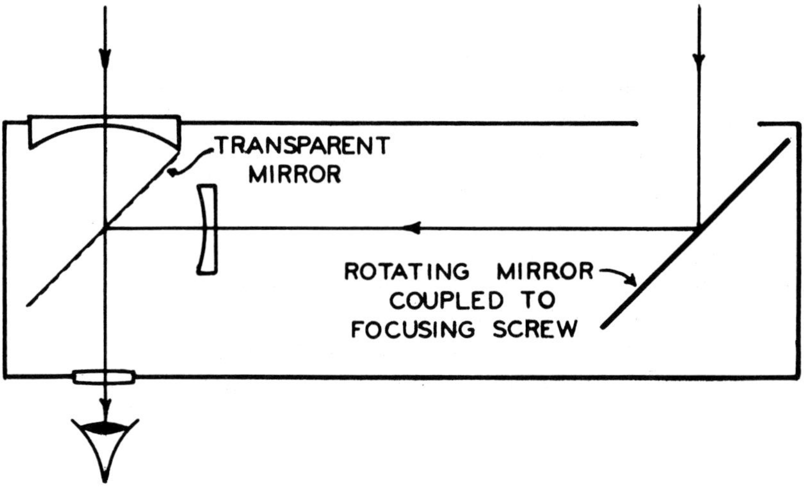

Figure 12.15. The principle of a combined viewfinder and rangefinder.

in Fig. 12.15. The partially reflecting mirror serves to reflect the rangefinder field into the middle of the viewfinder field, where it is superposed on the viewfinder image. By making the second image distinctively colored, no trouble in securing coincidence is ordinarily encountered.

The Single-Lens Reflex Camera

This form of camera was popular for many years in large sizes, typified in this country by the Graflex line. However, by about 1930 these cameras were losing popularity, chiefly because of their large size and weight, and the type survived eventually only in some European cameras such as the Korelle and Exakta. These early cameras were equipped with a focal-plane shutter, a 45° mirror behind the lens, a horizontal ground-glass focusing screen, and some arrangement for moving the mirror out of the way immediately before making an exposure. These early cameras were essentially waist-level devices.

In some reflex cameras, including the Retina Reflex, a between-lens shutter was used. The cycle of operations on pressing the release button was as follows:

The shutter closed.
The diaphragm dropped to its preset aperture.
The mirror lifted out of the way.

The shutter operated to make the exposure.
The mirror dropped back into the viewing position.
The iris opened fully.
The shutter opened.

The user then had to wind the film and cock the shutter ready for the next exposure.

Modern SLR Cameras

Soon after World War II some eye-level, single-lens reflex cameras began to make their appearance, using 35mm perforated film. At first a simple pentaprism was mounted over the ground-glass screen to permit eye-level operation, but the left-right image reversal and the fact that the image turned upside-down when the camera was rotated to permit photography of a tall object made this arrangement unsatisfactory. Starting with the Contax-S camera, and immediately followed by all manufacturers, the simple pentaprism was replaced by a roof pentaprism (Fig. 12.16), and the picture seen in the viewfinder then reproduced the scene exactly with no

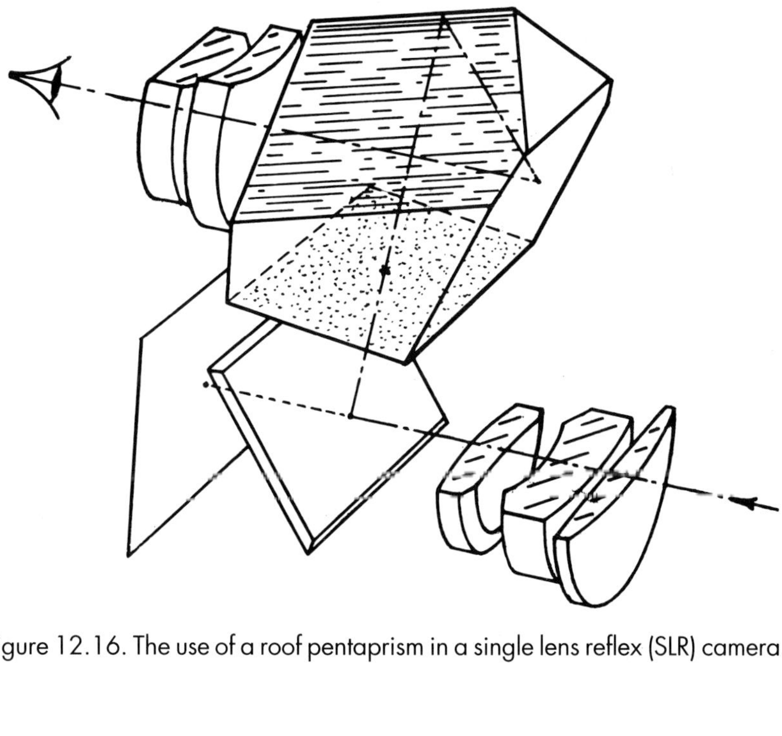

Figure 12.16. The use of a roof pentaprism in a single lens reflex (SLR) camera.

left-right inversion. If a 2-inch camera lens was used, and if the viewfinder lens had a focal length of 2 inches, the viewfinder constituted a unit-power telescope, giving a full-scale picture of the scene. Of course, if other interchangeable lenses were used, the scale of the viewfinder image would be altered accordingly, but the photographer always knew exactly how much of the scene would appear on the film.

Within the past 30 years this type of camera, the so-called SLR (single-lens reflex) has become almost universal among amateur photographers. Many additional features have been added, such as automatic exposure control, interchangeable lenses, a depth preview button, synchronized flash, a battery tester, a self-timer, coupled film transport and shutter cocking (to prevent unwanted double exposures), interchangeable focusing screens, etc., so that all the user has to do is select the subject and focus the camera. Even focusing is now automated in many SLR cameras. A very broad range of 35mm films are available, including black-and-white and color films, packaged in the standard metal cassette originally developed for the Retina camera line.

The cycle of operations is simpler if a focal-plane shutter is used. On pressing the release button the following sequence of actions occurs:

The mirror rises, blocking off stray light from the viewfinder.
The iris closes to its preset aperture.
The exposure is made.
The iris opens fully ready for the next exposure.
The mirror drops back into the viewing position.

The operator then cranks the film transport lever, which automatically cocks the shutter at the same time.

The modern SLR camera is the nearest thing to fully automatic photography. Some models have automatic exposure control, while in others it is necessary to match pointers to secure the correct exposure. With such a camera anything that can be seen can be photographed, even through a telescope or microscope, in good light or bad, close-up or distant, and by daylight or artificial light, as the occasion demands. With the availability of interchangeable and zoom lenses, the subject distance and the final image size are under complete control by the operator. Close-up diopter lenses, extension tubes and small bellows, and telephoto converters can be used without the necessity for calculations of image size, exposure, or focus setting. With a heavy camera and some support for the hands, exposures as long as one second can often be made hand-held, without the need for a tripod.

Automatic Focusing Devices

During the past twenty years several manufacturers have started to make cameras equipped with some means for automatic focusing. One may well question why this is necessary when visual focusing on a ground glass is so simple and rangefinders are so accurate, but the reason is that in some cases rapid focusing is required on a moving object. Sometimes the operator is unable to use the camera eyepiece because of obstructions or because the camera must be held high to avoid a crowd. Perhaps the operator is holding the camera by one hand and is unable to focus the lens manually, or perhaps a remote control arrangement is being used.

The history of automatic focusing somewhat resembles the history of exposure control. In the latter case, the first stage was eyeballing the scene. Then an external exposure meter was used, which was eventually coupled to the camera to indicate the correct exposure to the operator. Finally, the exposure control was made entirely automatic.

Focusing devices followed this evolution, except for the use of ground glass to provide easy visual focusing. In a camera lacking a ground-glass screen, an external rangefinder was often used; this was eventually coupled to the camera lens, and finally, the focusing operation has become entirely automatic.

One type of automatic focusing has been called *active*. Here, an infrared or acoustic pulse is emitted from the camera, and the time taken for the first reflection to arrive from objects in the scene is used to set the focus of the lens. This arrangement was used in the Polaroid Sonar camera of 1978. Unfortunately, if the closest object in the scene happens to be a window, the camera will focus on that! The automatic focusing can be overridden if this situation should arise.

The second type of automatic focusing is known as *passive*. For this system to be operative, the user must first orient the camera so that the object to be focused upon lies at the center of the field. Two images of this object are formed by two beams separated at the camera by the "rangefinder base" (see Fig. 12.17). In one beam the little 45° mirror is fixed, while in the other beam the mirror can be turned to cause the image to move laterally relative to the fixed image. The rotation of the movable mirror is coupled to the focusing mechanism in the camera.

In an ordinary visual rangefinder coupled to the focusing mechanism, the coincidence of the two images is readily observed. They may be mounted one above the other, or they may be superposed. For automatic focusing, each image is made to fall on a row of photodetectors, the signals from which are compared electronically. If the two images fall at corresponding positions on the rows of photodetectors, this indicates that the camera is in focus. If one image lies to the right or left of its correct position, a small motor causes the 45° mirror to rotate until the images are correctly situated, and the camera is then correctly focused. Several rangefinder cameras with this kind of automatic focusing appeared in the 1970s.

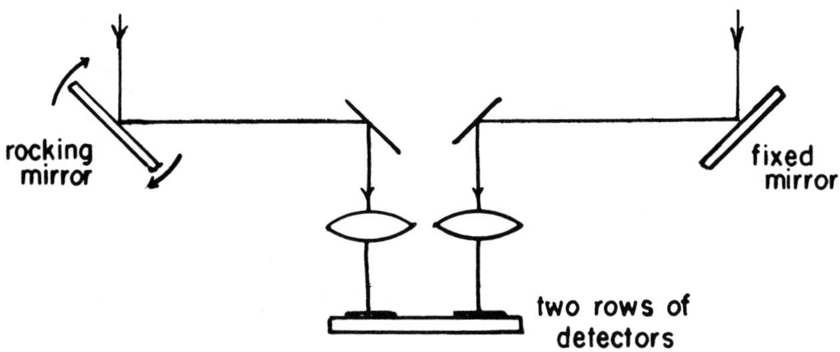

Figure 12.17. Automatic rangefinder focusing.

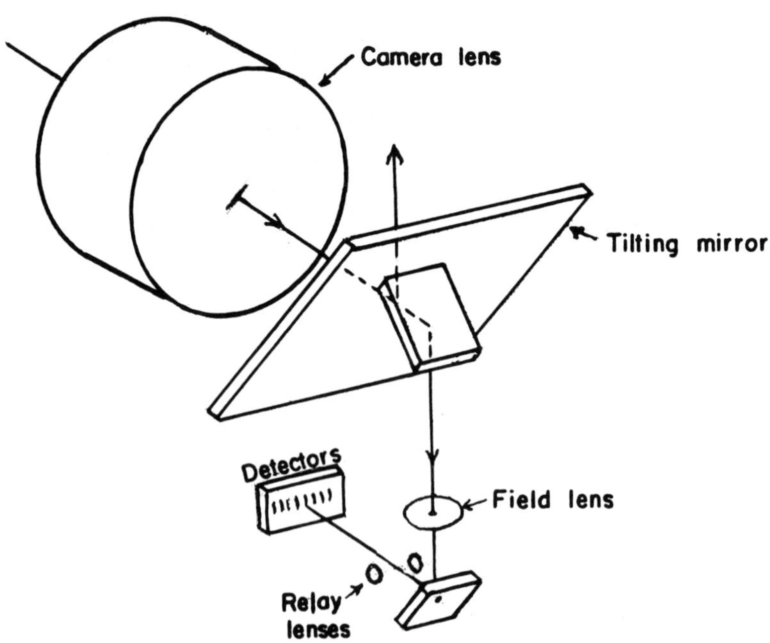

Figure 12.18. Automatic focusing in an SLR camera.

The application of automatic focusing to an SLR camera is much more difficult as there are not separate beams at the ends of a rangefinder base, and the user ordinarily has to judge focus by the sharpness of a ground-glass image. The "base" in this case is simply the diameter of the lens aperture.

The solution to the problem is shown in Fig. 12.18. A central area of the camera mirror is made partially transparent, and a tiny second "piggyback" mirror bends the beam downward to an image receiver with a small central hole and a field lens. After passing the field lens, another small 45° mirror turns the beam horizontally through a pair of tiny lenses that form images on two rows of photodetectors as in a rangefinder camera. The whole beam is only about 15° wide to function with an *f*/4.5 lens, so everything must be made and mounted with great precision. The process of automatic focusing is indicated in the straight-line diagram in Fig. 12.19, only one of the comparison beams being shown. Three situations are indicated. In (a) the camera lens is too far from the film, the image falling ahead of the field lens. The beam is then split between the two tiny lenses, which relay the image onto two rows of photodetectors. Here the images are shown too close together, and a small motor moves the lens closer to the film. In (b) the image falls in the desired film plane, and the images on the rows of detectors are at their correct relative positions. In (c) the lens is shown too close to the film, and the relayed images are now too far apart on the photodetectors.

At first, starting about 1980, the focusing motor was mounted in a "pod" on the side of the camera lens, but by 1985 the motor was placed

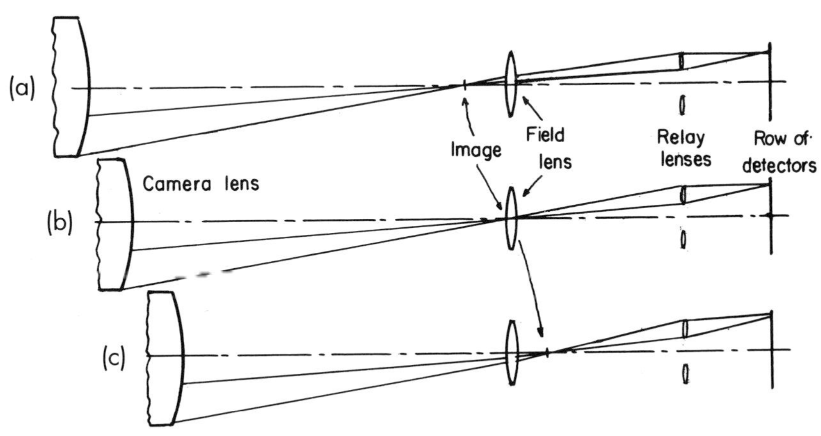

Figure 12.19. Automatic focusing arrangement for an SLR camera, straightened out.

inside the camera body, and the focusing means in the lens was coupled mechanically to a drive shaft, so that any suitably equipped lens could be used. Such lenses are engraved "AF," indicating automatic focusing. The body of a small lens can be moved for focusing, but larger lenses are focused by a movement of one or more elements inside the lens to reduce the load on the small motor.

Several excellent articles on the subject of automatic focusing have appeared in the literature.*

*L. A. Mannheim, "Autofocus: What's it all about, Mod. Phot. **44**, 102 (June 1980). S. F. Ray, *Applied Photographic Optics*, p. 173, Focal Press, London (1988). See also N. Goldberg, *Camera Technology: The Dark Side of the Lens*, Academic Press, New York (1992).

Index

N

O

P

R

S